LONDON MATHEMATICAL SOCIETY STUDENT TEXTS

Managing editor: Dr C.M. Series, Mathematics Institute
University of Warwick, Coventry CV4 7AL, United Kingdom

London Mathematical Society Student Texts 27

Hilbert Space: Compact Operators and the Trace Theorem

J.R. Retherford
Louisiana State University

CAMBRIDGE
UNIVERSITY PRESS

Published by the Press Syndicate of the University of Cambridge
The Pitt Building, Trumpington Street, Cambridge CB2 1RP
40 West 20th Street, New York, NY 10011-4211, USA
10 Stamford Road, Oakleigh, Melbourne 3166, Australia

First published 1993

Printed in Great Britain at the University Press, Cambridge

Library of Congress cataloging in publication data available

A catalogue record for this book is available from the British Library

ISBN 0 521 41884 4 hardback
ISBN 0 521 42933 1 paperback

TO:

Charles McArthur - Mathematics

Pat - Matrimony

Gunnar Johansen - Music*

*Gunnar Johansen, one of the great romantic pianists, and a friend for 35 years, died May 25, 1991 at the age of 85. Life is diminished by his passing.

ACKNOWLEDGEMENT

I would like to thank Natalie Wilson, Nell Castleberry, Loc Stewart, and Jacquie Rimes for the expert typing of my most difficult to read handwriting. I would also like to thank Mark Christie for his (sometimes) kind remarks concerning grammar and punctuation. Special thanks also to Susan Oncal for the drawings in Chapter Five.

Ron Retherford
Baton Rouge, LA

CONTENTS

x

INTRODUCTION

These chapters contain the material of a summer course (8 weeks) given at LSU a few years ago and repeated at Johannes Kepler Universität, Linz. In the summer the mathematics department at LSU is faced with offering courses that may be taken by graduate students at *all* levels: beginning to advanced Ph.D.! I hope that this material meets (and fills) that need.

For these lectures, the student will need a bit of mathematical sophistication and a fairly good course in advanced calculus (Cauchy Sequences, convergence of sequences, uniform continuity) and a good course in (finite dimensional) linear algebra (determinants, eigenvalues, linear transformations).

An undergraduate course in complex variables would also be nice. But, if the student was introduced to the line integral in calculus, the complex integration we do in these notes should present no difficulties. Knowledge of Lebesgue measure is *not* assumed. Thus, these notes will not discuss, e.g. $L_2[0,1]$ and thus also will not discuss integral operators given by L_2-kernels. (*To the student*: Forget this paragraph if it fails to make sense.)

Many will say that this omits too much from the theory of compact operators on Hilbert space. I claim not. It omits many important *examples* but in these notes we are interested in the *representation* of compact operators. From this point of view we have omitted nothing!

Our goal is to prove the Lidskij trace formula in as easy a fashion as possible. Another goal is to introduce the student to some Hilbert (and Banach) space techniques. After mastering these notes and a course in measure theory, the student should not have too much trouble absorbing some of the finer points of the general theory. Indeed, these notes have been written so that (with appropriate modifications, definitions, and a little effort) the main results can be generalized to certain classes of operators on general Banach spaces.

There are many exercises scattered through the text, as well as some "why" etc., in the proofs. The student is expected to do *all* the exercises and to try to understand *all* the proofs (answer the "why"!). There are a dozen important inequalities given throughout the text. To prove these inequalities, one often needs lesser inequalities. These easier inequalities have been assigned as exercises. Almost without exception these easier results all are "LaGrange Multiplier problems" and the student should refresh his memory on this subject (maximizing subject to a constraint). Credit, as best I know it, has been given for theorems and proofs.

At the end of each chapter, is a section entitled "Notes, exercises and hints." While the exercises in these sections are not necessary to achieve the proof of the Lidskij trace theorem, they do afford the opportunity for a better understanding of operators on Hilbert spaces and a chance to use some Banach space techniques. Extensive hints are given to keep the entire work user friendly. For those students desiring a challenge, cover up the hints.

These lectures have been influenced by many people, but primarily by H. König and A. Pietsch. I intended to write these lectures for publication a long time ago. I learned that both König and Pietsch were writing books on the subject. I, therefore, decided to wait and see what these books had to offer. These books are now in print (see Future Readings) - they have (as I suspected) much to offer! However, they are both written for the expert and not the novice. Thus I feel, even in this company, that these notes will still serve a useful function.

I chose to keep the references for this work at a minimum. I dropped a lot of names throughout the text. The books of Dunford and Schwartz (I and II) and Pietsch (Eigenvalues and s-numbers) contains hundreds of references concerning the material in these chapters. I recommend then to you (you won't find all the names!). The book of Pietsch also contains an amusing and detailed account of the (generalizations) of the material at hand.

Hopefully, the student will be intrigued by this survey of the the subject and will, at some point, study the books of König and Pietsch, and proceed to research.

0. THE INEQUALITIES OF IT ALL

1. Triangle Inequality: $\|x + y\| \le \|x\| + \|y\|$ (I.1c)

2. Cauchy-Schwarz-Bunyakovsky: $|(x, y)| \le (x, x)^{1/2}(y, y)^{1/2}$ (I.4)

3. Bessel: (u_i) orthonormal, $x \in H$, $n \ge 1$, $\sum_{i=1}^{n} |(x, u_i)|^2 \le \|x\|^2$ (II.2)

4. Riesz-Fischer: (two inequalities actually)

   (u_n) complete orthonormal, $\sum_{n=1}^{\infty} a_n u_n$ converges (in H) if and only if

   $$\sum_{n=1}^{\infty} |a_n|^2 < +\infty \qquad \text{(II.8)}$$

5. Parseval: (equality), (y_n) complete orthonormal, $x \in H$

   $$\|x\|^2 = \sum_{n=1}^{\infty} |(x, y_n)|^2 \qquad \text{(II.14)}$$

6. Holder: $\frac{1}{p} + \frac{1}{q} = 1$, $p \ge 1$ ($q = \infty$ if $p = 1$)

   $$\sum_{1}^{n} |a_i b_i| \le \left[\sum_{i=1}^{n} |a_i|^p \right]^{1/p} \left[\sum_{i=1}^{n} |b_i|^q \right]^{1/q}$$
   $$\left[\sum_{i=1}^{n} |a_i| \right] \max \{|b_i| : 1 \le i \le n\} \text{ if } q = \infty \qquad \text{(IX.2)}$$

7. Minkowski: $p \ge 1$,

   $$\left[\sum_{i=1}^{n} |a_i + b_i|^p \right]^{1/p} \le \left[\sum_{i=1}^{n} |a_i|^p \right]^{1/p} + \left[\sum_{i=1}^{n} |b_i|^p \right]^{1/p} \qquad \text{(IX.3)}$$

8. (weak) Weyl: $T \in K(H)$, $p \ge 1$

   $$\sum_{n=1}^{\infty} |\lambda_n(T)|^p \le \sum_{n=1}^{\infty} \sigma_n(T)^p \qquad \text{(IX.8)}$$

9.* Hadamard: (a_{ij}) an $n \times n$ matrix with values in $\mathbb{C}$

   $$|\det(a_{ij})| \le \prod_{j=1}^{n} \left(\sum_{i=1}^{n} |a_{ij}|^2 \right)^{1/2} \qquad \text{(Appendix B.1)}$$

10.* Weyl: for $T \in K(H)$, $|\lambda_1(T)| \ge |\lambda_2(T)| \ge \cdots$

$$\prod_{i=1}^{n} |\lambda_i(T)| \leq \prod_{i=1}^{n} \sigma_i(T) \qquad \text{(Appendix B.3)}$$

11. Hardy: $p > 1$, $a_i \geq 0$, $A_n = \sum\limits_{i=1}^{n} a_i$,

$$\sum_{n=1}^{m} \left(\frac{A_n}{n}\right)^p \leq \left(\frac{p}{p-1}\right)^p \sum_{n=1}^{m} a_n^p \qquad \text{(XI.3)}$$

12.* Comparison of geometric-arithemetic means: $a_i \geq 0$

$$\left(\prod_{i=1}^{n} a_i\right)^{1/n} \leq \frac{1}{n} \sum_{i=1}^{n} a_i \qquad \text{(Appendix C)}$$

(*) Appendices A, B and C can be omitted without any change to the remaining sections. The inequalities marked with * are not used elsewhere in the text. This, in no way, is meant to belittle their importance.

I. PRELIMINARIES

Exercise 1. Read the introduction!

The results we discuss in this monograph concern spectral properties of linear transformations on a very special space: *Hilbert Space.* This opening sentence should be meaningful before too many pages are turned!

The underlying space is a linear (= vector) space. Many of the results to follow are true for real or complex linear spaces but, to make the opening sentence meaningful we will usually assume throughout that the scalar field for our linear spaces is $\mathbb{C}$, the complex numbers.

Thus, let X be a linear space over $\mathbb{C}$. In this space we need a way to measure distance between elements.

I.1 Definition. A *norm*, $\| \bullet \|$, is a function from X into the non-negative reals $\mathbb{R}^+$ satisfying

(a) $\|x\| = 0$ if and only if $x = 0$; (the first "0" is the real number 0 and the second is the zero of the vector space X!)

(b) for each $x \in X$ and each scalar $\alpha \in \mathbb{C}$

$$\|\alpha x\| = |\alpha| \|x\|; \text{ and,}$$

(c) "the triangle inequality" $\|x + y\| \leq \|x\| + \|y\|$ for all $x, y \in X$.

We emphasize that, by definition, $\|x\| \geq 0$ for all $x \in X$.

Let's consider a few norms on *real* spaces to see that the concept of norm generalizes the notion of absolute value, and more generally the notion of the length of a vector.

Some Examples:

(a) Let $\mathbb{R}_1$ denote the real line with the usual arithmetic. The usual absolute value, $|x|$, is a norm.

(b) Consider $\mathbb{R}_n$, the real space of n-tuples $x = (x_1, x_2, \ldots, x_n)$. Let

$$\|x\|_2 = \left(\sum_{i=1}^{n} x_i^2 \right)^{1/2}.$$

Then $\| \bullet \|$ is a norm. (We'll have plenty to say about this later).

(c) Define $\|x\|_1 = \sum\limits_{i=1}^{n} |x_i|$ on $\mathbb{R}^n$. Then $\| \bullet \|_1$ is a norm.

(d) Define $\|x\|_\infty = \max(|x_1|, \ldots, |x_n|)$ on $\mathbb{R}_n$. Then $\| \bullet \|_\infty$ is a norm.

The student should check these statements. Similarly, we can define norms on $\mathbb{C}$ and $\mathbb{C}_n$, the space of complex n-tuples. Let us look at a less familiar space.

(e) Let $C[a, b]$ denote the space of real (or complex) valued continuous functions on the interval $[a, b]$. For $f \in C[a, b]$ define $\|f\| = \sup\{|f(t)| : t \in [a, b]\}$. Then $\| \bullet \|$ is a norm on $C[a, b]$. (Here you must use the fact that a continuous real-valued function on $[a, b]$ achieves its maximum).

If one has a linear space X and a norm $\| \bullet \|$ defined on X, define the distance between $x, y \in X$ by $\|x - y\|$. Once we have this distance function given by a norm, we can extend familiar concepts from the calculus to this more general setting.

I.2 Definition. Let (x_n) be a sequence in X and $\| \bullet \|$ a norm on X. We say that (x_n) is a *Cauchy sequence* (with respect to the given norm) if for every $\varepsilon > 0$ there is an integer N such that $m, n \geq N$ implies $\|x_n - x_m\| < \varepsilon$. We say that the sequence (x_n) has a *limit* x in X (or that (x_n) *converges* to x) in the given norm provided that for every $\varepsilon > 0$ there is an integer N such that $n \geq N$ implies $\|x_n - x\| < \varepsilon$. We write $\lim\limits_{n \to \infty} \|x_n - x\| = 0$ or $\lim\limits_{n \to \infty} x_n = x$. A function defined on X is *continuous* at $x \in X$ provided that for every sequence (x_n) in X converging to x, the sequence $(f(x_n))$ converges to $f(x)$. Here, for convenience, the range of f is assumed to be in a vector space equipped with a norm (although this really isn't essential nor all together desirable).

If "+" and "•" are the operations of addition and scalar multiplication on X, these operations are continuous in the following sense: if $\lim\limits_{n \to \infty} x_n = x$ and $\lim\limits_{n \to \infty} y_n = y$ then $\lim\limits_{n \to \infty} x_n + y_n = x + y$. If $\lim\limits_{n \to \infty} \alpha_n = \alpha$ (in $\mathbb{C}$) then $\lim\limits_{n \to \infty} \alpha_n x_n = \alpha x$. These facts result from the triangle inequality and should be proved by the student.

A linear space X equipped with a norm $\| \bullet \|$ and distance between vectors measured by this norm is, not surprisingly, called a *normed linear space*. If every Cauchy sequence in a normed linear space X has a limit in X then X is said to be *complete*. A complete normed linear space is called a *Banach* space after the great Polish mathematician Stefan Banach. The Hilbert spaces we are after are Banach spaces whose norms have very special properties.

I.3 Definition. Let X be a linear space over $\mathbb{C}$. An *inner -product*, $(\bullet, \bullet)$, is a function defined on $X \times X$, the set of all pairs of elements in X, satisfying

(a) $(x, x) \geq 0$;

(b) $(x, x) = 0$ if and only if $x = 0$;

(c) $(\alpha x, y) = \alpha(x, y)$ for all $x, y \in X$ and $\alpha \in \mathbb{C}$;

(d) $(x, \alpha y) = \overline{\alpha}(x, y)$ for all $x, y \in X$, $\alpha \in \mathbb{C}$. Here $\overline{\alpha}$ denotes the complex conjugate of α; we will call this property *conjugate homogeneity*.

(e) $(x, y) = (\overline{y, x})$; and

(f) $(x_1 + x_2, y) = (x_1 + y) + (x_2, y)$ for all $x_1, x_2, y \in X$.

For example in $\mathbb{C}_n$, $(x, y) = \sum_{i=1}^{n} x_i \overline{y}_i$ is an inner-product. Of course, the student should check this!

One of the most important inequalities in mathematics concerns inner-products.

I.4 Theorem. (**Cauchy-Schwarz-Bunyakovsky Inequality**) If $(\bullet, \bullet)$ is an inner-product on a linear space X then for all $x, y \in X$

$$|(x, y)| \leq (x, x)^{1/2}(y, y)^{1/2}.$$

Proof: For $x, y \in X$ let $b = \frac{(x, y)}{|(x, y)|}$ if $(x, y) \neq 0$ and $b = 1$ if $(x, y) = 0$; also, let a be an arbitrary real number. Using the properties of an inner-product, we obtain

$$(ax + by, ax + by) = a^2(x, x) + a\overline{b}(x, y) + ba(y, x) + b\overline{b}(y, y) \geq 0.$$

Now, using the definition of b and $(x, y) = (\overline{y, x})$, we obtain

(*) $\qquad\qquad a^2(x, x) + 2a|(x, y)| + (y, y) \geq 0.$

If $x = 0$ there is nothing to prove and if $x \neq 0$ (*) has a minimum at $a = \frac{-|(x, y)|}{(x, x)}$ (just take the derivative as a function of a). Putting this value of a into (*) yields the inequality.

One of our main applications of the Cauchy-Schwarz-Bunyakovsky inequality is the following:

I.5 Theorem.   If $(\bullet, \bullet)$ is an inner-product on $X \times X$ then $(x, x)^{1/2}$ is a norm on X.

Proof: Everything in the definition of a norm follows from the corresponding properties of the inner-product except the triangle inequality. For $x, y \in X$ (writing $\|x\| = (x, x)^{1/2}$) we have

$$
\begin{aligned}
\|(x + y)\|^2 &= (x, x) + (x, y) + (y, x) + (y, y) \\
&= \|x\|^2 + 2\mathrm{Re}(x, y) + \|y\|^2 \\
&\leq \|x\|^2 + 2|(x, y)| + \|y\|^2 \\
&\leq \|x\|^2 + 2\|x\|\|y\| + \|y\|^2 = (\|x\| + \|y\|)^2.
\end{aligned}
$$

The last inequality is just the Cauchy-Schwarz-Bunyakovsky inequality. The symbol $\mathrm{Re}z$ denotes the real part of $z \in \mathbb{C}$.

You should now check that on $\mathbb{R}_n$ or $\mathbb{C}_n$

$$
\|x\| = \left(\sum_{i=1}^{n} |x_i|^2 \right)^{1/2}
$$

is a *norm* coming from the *inner-product*

$$
(x, y) = \sum_{i=1}^{n} x_i \bar{y}_i.
$$

A Banach space whose norm $\| \bullet \|$ comes from an inner-product via $\|x\| = (x, x)^{1/2}$ is called a *Hilbert-space* in honor of the great German mathematician David Hilbert.

Exercise 2.   If $\| \bullet \|$ is such a norm, prove the parallelogram law: $\|x + y\|^2 = 2 \left[\|x\|^2 + \|y\|^2 \right]$.

A normed linear space (not complete) is called a *pre-Hilbert space* if its norm comes from an inner-product in this fashion. Hilbert and pre-Hilbert spaces are called inner-product spaces.

We can now present the main example of Hilbert spaces (at least for our purposes!).

I.6 Example. The space ℓ_2.

Let $\ell_2 = \left\{ (a_n) \mid \sum_{n=1}^{\infty} |a_n|^2 < +\infty \right\}$ with coordinate-wise arithemetic. Here the (a_n) are sequences of complex numbers (although at this stage they could be real sequences). That ℓ_2 is a linear space follows from $ab \leq \frac{1}{2} \left[a^2 + b^2 \right]$. If $a = (a_n)$, $b = (b_n) \in \ell_2$ then $(a, b) = \sum_{n=1}^{\infty} a_n \bar{b}_n$ is an inner product. Indeed $(|a_n|)$ and $(|b_n|)$ are in ℓ_2 and for fixed N the Cauchy-Schwarz-Bunyakovsky inequality gives

$$\sum_{n=1}^{N} |a_n||\bar{b}_n| \leq \left(\sum_{n=1}^{N} |a_n|^2 \right)^{1/2} \left(\sum_{n=1}^{N} |b_n|^2 \right)^{1/2}$$

$$\leq \left(\sum_{n=1}^{\infty} |a_n|^2 \right)^{1/2} \left(\sum_{n=1}^{\infty} |b_n|^2 \right)^{1/2}.$$

Letting $N \to \infty$ yields

$$\sum_{n=1}^{\infty} |a_n||b_n| \leq \left(\sum_{n=1}^{\infty} |a_n|^2 \right)^{1/2} \left(\sum_{n=1}^{\infty} |b_n|^2 \right)^{1/2}$$

so (a, b) is well-defined. It is routine to check that (a, b) has all the properties of an inner-product. In particular observe that

$$\left(\sum_{n=1}^{\infty} |a_n|^2 \right)^{1/2} = (a, a)^{1/2}$$

is the inner-product norm on ℓ_2.

With this norm ℓ_2 is complete! This fact is non-trivial but a proof should be attempted by the student (perhaps with a lot of help from the instructor). Thus ℓ_2 is a Hilbert space. Let e_n denote the sequence with 1 the nth term and the remaining terms 0. Clearly each $e_n \in \ell_2$ and any finite set $\{e_1, e_2, \ldots, e_N\}$ is linearly independent, so ℓ_2 is *not* a finite-dimensional linear space!

From now on H will denote a Hilbert space. We should remark that the inner-product $(\bullet, \bullet)$ on $H \times H$ is continuous in the following sense: If $\lim_{n \to \infty} x_n = x$ and $\lim_{n \to \infty} y_n = y$ then $\lim_{n \to \infty} (x_n, y_n) = (x, y)$. Clearly, the statement $\lim_{n \to \infty} (x_n, y_n) = (x, y)$ is equivalent to $\lim_{n \to \infty} [(x_n - x, y_n) + (x, y_n - y)] = 0$ (why?) and this is immediate from the Cauchy-Schwarz-Bunyakovsky inequality. (You need the fact that a Cauchy sequence is bounded:

$\sup_n \|y_n\| \leq M$ for some $M > 0$.) We will now begin studying the special properties enjoyed *only* by Hilbert spaces among the Banach spaces.

Remarks, Exercises and Hints

The axioms for finite dimensional Euclidean Spaces (our spaces $\mathbb{R}_n(\mathbb{C}_n)$) were given by Herman Weyl about 1910. Hilbert spaces were studied by David Hilbert and his Tübingen school but not in an abstract fashion. The axioms for what we now call Hilbert space were given by John VonNeumann in 1927. In 1932 Stefan Banach introduced the axioms for what he called "spaces of type B". Mathematicians took the hint and thus Banach spaces came into being. Earlier, similiar ideas had been used by Norbert Wiener and T. H. Hildebrandt. In the older literature you will find Hildebrandt never yielded and referred to Banach spaces as NLCS(normed, linear complete spaces)!

The notion of inner-product (or dot product or scalar product) goes back to the early days of vector analysis.

The Cauchy-Schwarz-Bunyakovsky inequality was proved for finite sums by Cauchy (1821), for integrals by Bunyakovsky (1859) and rediscovered by Schwarz in 1885. We will see that the Cauchy-Schwarz-Bunyakovsky inequality is a special case of Hölders inequality.

1. In $\mathbb{R}_2$, view (a,b) as the directed segment (vector) from (0,0) to (a,b). Given two vectors u and v let Θ be the angle between them satisfying $0 \le \Theta \le \pi$.

   a. Prove that $(u,v) = \|u\|\|v\| \cos\Theta$ and conclude that Θ is acute (obtuse) if $(u,v) > 0\big((u,v) < 0\big)$ and $\Theta = \frac{\pi}{2}$ if and only if $(u,v) = 0$.

2. In definition I.3, (d) is superfulous.

3. Show that an inner product is additive in the second variable:
$$(x, y_1 + y_2) = (x, y_1) + (x, y_2) \text{ for all } x, y_1, y_2.$$

4. If X is an inner product space and $(x,u) = (x,v)$ for all $x \in X$ then $u = v$.

5. If X is a real inner product space, show that $(x - y, x + y) = 0$ if and only if $\|x\| = \|y\|$.

6. If X is an inner product space and $\lim_{n\to\infty}(x_n, x) = (x, x)$ and $\lim_{n\to\infty} \|x_n\| = \|x\|$ then $\lim_{n\to\infty} x_n = x$.

7. Do the norms $||x||_1 and ||x||_\infty$ defined on $\mathbb{R}_n, n > 2$ come from an inner-product? What about $n = 1$?

8. Answer problem 7 for the space $C[a, b]$ with the supremum norm.

Finally a more difficult problem.

In Exercise 2 of this chapter you were asked to prove the parallelgram law from inner-product spaces.

9. (Jordan, VonNeumann) Prove that if X is a normed linear space with a norm satisfying (*) $||x + y||^2 + ||x - y||^2 = 2[||x||^2 + ||y||^2]$ the norm comes from an inner-product.

    [HINT] If X is real, define

$$(x, y) = \frac{1}{4}[||x + y||^2 - ||x - y||^2];$$

if X is complex, define

$$Re(x, y) = \frac{1}{4}[||x + y||^2 - ||x - y||^2] \text{ and } Im(x, y) = \frac{1}{4}[||x + iy||^2 - ||x - iy||]$$

(Rez and Imz denote the real and imaginary parts of the complex number z). To prove (+)$(\alpha x, y) = \alpha(x, y)$ use induction to prove (+) for $\alpha = \frac{1}{n}$. Then, use induction again to prove (+) for $\frac{m}{n}$. If α is irrational write $\alpha = \lim_{n \to \infty} r_n$ with r_n rational and invoke a continuity argument. For the complex case use the above ideas applied to the real and imaginary parts of (x, y).

For the remainder of the proof use the following big hint of M.M. Day: In the parallelogram equality replace x by $x \pm z$ obtaining

$$||x + z + y||^2 + ||x + z - y||^2 = ||x - z + y||^2 - ||x - z - y||^2$$
$$= 2[||x + z||^2 + ||y||^2 - ||x - z||^2 - ||y||^2].$$

Use this to show that

$$4(x + y, z) + 4(x - y, z) = 8(x, z).$$

Letting $x + y = a \qquad x - y = b$ yields

$$(a, z) + (b, z) = (a + b, z).$$

II. ORTHOGONALITY

Let X be an inner-product space. The main property distinguishing X from general normed linear spaces is that of *orthogonality*. Various orthogonality conditions have been introduced into normed linear spaces, notably by R. C. James. The next definition however, only makes sense in inner-product spaces.

II.1 Definition. Let $x, y \in X$, an inner-product space. We say that x *is orthogonal* to y, and write $x \perp y$ provided the inner-product $(x, y) = 0$. If S is a set in X we say that S is an *orthogonal* set if $x, y \in S$ and $x \neq y$ implies $(x, y) = 0$. If each element of an orthogonal set S has norm 1 we say that S is an *orthonormal* set.

This may be the most important concept in Hilbert spaces.

Let us remark that if $x \perp y$ then $y \perp x$. Also $x \perp 0$ for $x \in X$. If $\{u_1, u_2, \ldots, u_n\}$ is an orthonormal set in X then $(u_i, u_j) = \delta_{ij}$, the Kronecker delta (which is 1 if $i = j$ and 0 otherwise).

Next we give another famous inequality which will be of utmost importance later on.

II.2 Theorem. (**Bessel's Inequality**). Let S be an orthonormal set in an inner-product space X. Let $\{u_1, \ldots, u_n\}$ be a finite subset of S and $x \in X$. Then

$$\sum_{i=1}^{n} |(x, u_i)|^2 \leq \|x\|^2.$$

Proof: Compute $\left(x - \sum_{i=1}^{n}(x, u_i)u_i, x - \sum_{i=1}^{n}(x, u_i)u_i \right)$ with the observation that, by orthogonality

$$\left(\sum_{i=1}^{n}(x, u_i)u_i, \sum_{j=1}^{n}(x, u_j)u_j \right) = \sum_{i,j=1}^{n}(x, u_i)(\overline{x, u_j})(u_i, u_j) = \sum_{i=1}^{n} |(x, u_i)|^2.$$

We have

$$0 \le \left(x - \sum_{i=1}^{n}(x, u_i)u_i, x - \sum_{i=1}^{n}(x, u_i)u_i \right)$$

$$= \|x\|^2 - \left(\sum_{i=1}^{n}(x, u_i)u_i, x \right) - \left(x, \sum_{j=1}^{n}(x, u_j)u_j \right) + \sum_{i=1}^{n}|(x, u_j)|^2$$

$$= \|x\|^2 - \sum_{i=1}^{n}|(x, u_i)|^2.$$

Recall that a set S is *countable* (or *countably infinite*) if there is a one-to-one function from the positive integers *onto* S. Intuitively, we can, in this case, enumerate the elements of S as a sequence: $S = \{s_1, s_2, s_3, \ldots\}$.

A set that is finite or countable is said to be *at most countable* . We need the fact that the countable union of finite or countable sets is again at most countable. The introduction to almost any advanced calculus course contains this result. For completeness sake we sketch the proof here: First observe that a subset of a countable set is at most countable (prove this!). Thus to show that a set is at most countable it suffices to show that there is a one-to-one function from that set *into* the positive integers ω.

First we show that $\omega \times \omega$ is at most countable. Define, e.g. $f(m, n)$ on $\omega \times \omega$ by $f(m, n) = 2^m 3^n$ if $(m, n) \in \omega \times \omega$. Since 2,3 are relatively prime (i.e. have no common divisors except 1) f is one-to-one, and $\omega \times \omega$ is (in fact) countable. Next we show that if (S_n) is an at most countable collection of disjoint at most countable sets, then the union

$$S = \bigcup_{n=1}^{\infty} S_n$$

is at most countable (sounds like double talk doesn't it?). To see this, since each S_n is at most coutable we can (via the one-to-one mapping to ω) enumerate $S_n = \{s_{1,n}, s_{2,n}, s_{3,n}, \ldots\}$ for $n = 1, 2, \ldots$. If $s \in S$ then $s \in S_n$ for a unique n since the S_n are pairwise disjoint ($S_n \cap S_m = \emptyset$) thus $s = s_{k,n}$ and k is also uniquely determined (why?). Thus if f is defined on S by $f(s) = (k, n)$ then f is a one-to-one function from S *into* $\omega \times \omega$ and so is at most countable. For the last step observe that if the sets (S_n) are not pairwise disjoint they can be replaced by pairwise disjoint sets (T_n)

where $T_n \subset S_n$ and $\bigcup\limits_{n=1}^{\infty} T_n = \bigcup\limits_{n=1}^{\infty} S_n$. Indeed, let $T_1 = S_1$ and for $n > 1$ let $T_n = S_n \setminus \bigcup\limits_{k=1}^{n-1} S_k$. Putting all the above together shows that the (at most) countable union of (at most) countable sets is (at most) countable. If some of the sets are finite a very slight modification is needed above. We leave this to the student.

II.3 Corollary to Bessel's inequality. Let S be an orthonormal set in X and for let $x \in X$ let $A(x) = \{s \in S | (x, s) \neq 0\}$. Then $A(x)$ is at most countable.

Proof: If S is at most countable there is nothing to prove. If not, then for any k we have by Bessel's inequality

$$\sum_{i=1}^{k} |(x, u_i)|^2 \leq \|x\|^2 \quad \text{for} \quad u_1, \ldots, u_k \in S.$$

If $A_n(x) = \{s \in A(x) | |(x, s)| \geq \frac{1}{n}\}$ and $u_1, \ldots, u_k \in A_n(x)$ we have

$$\frac{k}{n^2} \leq \sum_{i=1}^{k} |(x, u_i)|^2 \leq \|x\|^2$$

and so $A_n(x)$ is finite. Clearly $A(x) = \bigcup\limits_{n=1}^{\infty} A_n(x)$.

Using Corollary II.3, it is possible to give meaning to the symbol

$$\sum_{u \in S} (x, u) u$$

where S is an orthonormal set and $x \in X$. Indeed, since all but a countable number of the terms $(x, u)u$ are 0 by II.3, we should have

$$\sum_{u \in S} (x, u) u = \sum_{i=1}^{\infty} (x, u_i) u_i$$

for suitable (u_i) in S. However there are problems with this: the (u_i) change with x and no order has been specified for the (u_i). We do not know how they sit in S. Thus to attach unambiguous meaning to $\sum\limits_{i=1}^{\infty} (x, u_i) u_i$ we must insist that no matter how the non-zero terms $(x, u)u$ are arranged, the corresponding series converges. This type of convergence is called *unconditional convergence.* Formally, a series $\sum_{n=1}^{\infty} y_n$ in X is unconditionally convergent if the series $\sum_{n=1}^{\infty} y_{\tau(n)}$ converges for every one-to-one function mapping the positive integers onto themselves (permutation). There is still a problem! We must show that all of these re-arranged series necessarily converge to the same element.

If Σy_n is unconditionally convergent in X and σ, τ are two permutations of the positive integers let $x = \sum_{n=1}^{\infty} y_{\tau(n)}$ and $z = \sum_{n=1}^{\infty} y_{\sigma(n)}$. Since $\sum_{n=1}^{\infty} (y_{\rho(n)}, y)$ converges for each permutation ρ and each $y \in X$ (by the continuity of the inner-product) it follows that $\sum_{n=1}^{\infty} |(y_n, y)|$ converges (why?). By the classical theorem of Riemann (all re-arrangements of an absolutely convergent series converge to the same value), it follows that

$$0 = \sum_{n=1}^{\infty} (y_{\tau(n)} - y_{\sigma(n)}, x - z) = \sum_{n=1}^{\infty} (y_{\tau(n)}, x) - \sum_{n=1}^{\infty} (y_{\sigma(n)}, x)$$
$$- \sum_{n=1}^{\infty} (y_{\tau(n)}, z) + \sum_{n=1}^{\infty} (y_{\sigma(n)}, z) = \|x\|^2 - (z, x) - (x, z) + \|z\|^2$$
$$= \|x - z\|^2,$$

i.e., $x = z$. Thus our definition yields a unique sum.

It is terribly old-fashioned (and unnecessary if the student completely understands the above) but we mention that unconditional convergence can be phrased in a manner completely analogous to that of ordinary (ordered) convergence: Let Σ denote the finite subsets of the positive integers. A series Σy_n in X is *Moore-Smith convergent* to x if for $\varepsilon > 0$ there is an $\sigma_0 \in \Sigma$ such that for $\sigma \supset \sigma_0$, $\sigma \in \Sigma$,

$$\left\| \sum_{i \in \sigma} x_i - x \right\| < \varepsilon.$$

The student should compare this with the definition of *convergent*. It is fairly easy to see that unconditional convergence and Moore-Smith convergence are the same! Indeed if $\sum_{n=1}^{\infty} x_n$ is Moore-Smith convergent to x then for $\varepsilon > 0$ there is a $\sigma_0 \in \Sigma$ such that $\left\| \sum_{i \in \sigma} x_i - x \right\| < \varepsilon$ whenever

$\sigma_0 \subset \sigma$ and $\sigma \in \Sigma$. Let τ be a permutation of the positive integers and $\sigma' = \tau^{-1}(\sigma_0)$. Let $n = \max\{i : i \in \sigma'\}$. If $p > n$, $\sigma' \subset \{1, 2, \ldots, p\}$ and so $\sigma_0 = \tau(\sigma') \subset \tau\{1, 2, \ldots, p\}$. Thus,

$$\left\| \sum_{i=1}^{P} x_{\tau(i)} - x \right\| < \varepsilon, \quad \text{i.e.} \quad \sum_{n=1}^{\infty} x_n$$

is unconditionally convergent. If, on the other hand $\sum_{n=1}^{\infty} x_n$ is unconditionally convergent we have shown that all rearrangements have the same sum, say x. If $\sum_{n=1}^{\infty} x_n$ is not Moore-Smith convergent to x there is an $\varepsilon > 0$ such that for all $\sigma \in \Sigma$ there is a $\sigma' \in \Sigma$, $\sigma' \supset \sigma$ such that $\left\| \sum_{i\in\sigma'} x_i - x \right\| \geq \varepsilon$. Choose m_1 such that $n \geq m_1$ implies $\left\| \sum_{i=1}^{n} x_i - x \right\| < \varepsilon$ and let $\sigma_1 = \{1, 2, \ldots, m_1\}$. Choose $\sigma_2 \supset \{1, 2, \ldots, m_1\}$ with $\left\| \sum_{i\in\sigma_2} x_i - x \right\| \geq \varepsilon$. Choose $\sigma_3 = \{1, \ldots, m_3\} \supset \sigma_2$ such that $\left\| \sum_{i\in\sigma_3} x_i - x \right\| < \varepsilon$. In general choose $\sigma_n \subset \sigma_{n+1}$, $\sigma_{2k-1} = \{1, 2, \ldots, m_{2k-1}\}$, $\left\| \sum_{i\in\sigma_{2k-1}} x_i - x \right\| < \varepsilon$ and $\left\| \sum_{i\in\sigma_{2k}} x_i - x \right\| \geq \varepsilon$. Let τ be the permutation determined by $\sigma_1, \sigma_2 \backslash \sigma_1, \sigma_3 \backslash \sigma_2, \ldots$. Clearly $\sum_{i=1}^{\infty} x_{\tau(i)}$ does not converge.

This digression is just to give a feeling for unconditional convergence. The important point is that the *order of summation does not matter*.

After all this work we need examples to show that an orthonormal set is not necessarily countable.

We generalize the space ℓ_2 to give such examples.

Let Γ be an arbitrary set and let $\ell_2(\Gamma) = \{(\alpha_\gamma)_{\gamma\in\Gamma} | \sum_{\gamma\in\Gamma} |\alpha_\gamma|^2 < +\infty\}$ with the meaning that for each $a = (a_\gamma)$ only at most countably many $a_\gamma \neq 0$. Thus, ℓ_2 is $\ell_2(\omega)$, ω the positive integers. With the meaning of convergence of series as given above it is easy to see that for $a = (a_\gamma)$, $b = (b_\gamma) \in \ell_2(\Gamma)$, $(a, b) = \sum_{\gamma\in\Gamma} a_\gamma \bar{b}_\gamma$ is an inner-product. If e_γ is the function on Γ which is 1 at γ and 0 otherwise, it is trivial to check that $\{e_\gamma\}$ is an orthonormal set. If Γ is not countable (e.g., $\mathbb{R}_1$ or $\mathbb{C}$) then $\{e_\gamma\}$ is obviously not countable.

Luckily for us, we generally need only work with countable sets. However, continuing in the above vein let's give one more result which is immediate from the Cauchy-Schwarz-Bunyakovsky and Bessel inequalities.

II.4 Corollary. Let S be an orthonormal set in X and $x, y \in X$. Then

$$\sum_{u \in S} |(x, u)(\overline{y, u})| \leq \|x\| \|y\|.$$

Proof: All but countably many of the terms on the left-hand side of the inequality are zero, so

$$\sum_{u \in S} |(x, u)(\overline{y, u})| = \sum_{i=1}^{\infty} |(x, u_i)(\overline{y, u_i})|$$

for some $\{u_i\}$ in S. For fixed n

$$\sum_{i=1}^{n} |(x, u_i)(\overline{y, u_i})| \leq \left(\sum_{i=1}^{n} |(x, u_i)|^2 \right)^{1/2} \left(\sum_{i=1}^{n} |(y, u_i)|^2 \right)^{1/2}$$

by the Cauchy-Schwarz-Bunyakovsky inequality for ℓ_2. The right hand side of this expression is $\leq \|x\| \|y\|$ by Bessel's inequality.

Before proceeding we borrow some terms from topology.

II.5 Definition. Let $B \subset A \subset X$. By the *closure* of A, written $\overline{A}$, we mean the collection of all $x \in X$ such that there is a sequence (x_n) in A with $\lim_{n \to \infty} x_n = x$. A set A is *closed* in X if $A = \overline{A}$. The set $B \subset A$ is *dense* in A if for each $\varepsilon > 0$ and $a \in A$ there is a $b \in B$ with $\|a - b\| < \varepsilon$.

For instance the rational numbers are dense in the reals, and the complex numbers of the form $a + bi$, a, b rational, are dense in $\mathbb{C}$ (check!).

II.6 Definition. A normed linear space Y is *separable* if Y contains a countable dense set.

It is an important fact that ℓ_2 is separable. In fact if B is the set of those sequences in ℓ_2 which have only finitely many non-zero terms and the non-zero entries are of the form $a + bi$, a, b rational, then B is countable and $\overline{B} = \ell_2$. The student should prove these statements (re-examine the earlier material on countable sets!).

How large can an orthonormal set be in a separable inner-product space?

II.7 Theorem. If X is separable and S an orthonormal set in X then S is at most countable.

Proof: If S is finite there is nothing to prove. If S is infinite it suffices to exhibit a one-to-one function from S *into* ω, the positive integers. First observe that if $x, y \in S$, $\|x - y\| = \sqrt{2}$ (since they are orthonormal). Let (y_n) be a countable dense set in X. Choose m, n so that

$$\|x - y_n\| < \frac{\sqrt{2}}{4} \quad \text{and} \quad \|y - y_m\| < \frac{\sqrt{2}}{4}.$$

Then $\sqrt{2} = \|x - y\| = \|x - y_n + y_n - y_m + y_m - y\| \leq \|y_n - y_m\| + \frac{\sqrt{2}}{2}$ so $y_m \neq y_n$. In particular $m \neq n$. For $x \in X$ let $f(x) = n$ as above.

We are now able to prove one of the most important theorems in Hilbert space theory (at last we're talking about Hilbert spaces!).

II.8 The Riesz-Fischer Theorem. Let H be separable Hilbert space and $\{u_n\}$ an orthonormal set in H. Then $\sum\limits_{n=1}^{\infty} a_n u_n$ converges if and only if

$$\sum_{n=1}^{\infty} |a_n|^2 < +\infty \quad [\text{i.e. } (a_n) \in \ell_2].$$

In such a case, if $x = \sum\limits_{n=1}^{\infty} a_n u_n$ then $a_n = (x, u_n)$.

Proof: Since H is complete and ℓ_2 is complete (exercise) we need only observe that appropriate partial sums are Cauchy. Thus let $S_n = \sum\limits_{i=1}^{n} a_i u_i$ and suppose also that $n > m$. Then

$$\|S_n - S_m\|^2 = \left\| \sum_{m+1}^{n} a_i u_i \right\|^2 = \sum_{m+1}^{n} |a_i|^2 \quad (\text{why?})$$

i.e., $\left(\sum\limits_{i=1}^{n} |a_i|^2 \right)$ is Cauchy if and only if (S_n) is Cauchy.

The last statement follows from the continuity of the inner product: if

$$x = \sum_{n=1}^{\infty} a_n u_n, \ a_j = \lim_{n \to \infty} \left(\sum_{i=1}^{n} (a_i u_i, u_j) \right) = (x, u_j).$$

This result may appear trivial, but discovering the proper setting was not easy.

If $x \in X$ (inner product space) and A is a subset of X we write $x \perp A$ if $(x,a) = 0$ for all $a \in A$. We write sp A, the span of A, for the set of all finite linear combinations of elements of A. Clearly, by the linearity and continuity of $(\bullet, \bullet)$, if $x \perp A$ then $x \perp \overline{\text{sp } A}$ (check!).

II.9 Theorem.   (Beppo Levi, I think). Let H be a Hilbert space and S an orthonormal set in H. For each $x \in H$ let $x(S)$ denote the formal sum $\sum_{u \in S} (x,u)u$. Then $x \in \overline{\text{sp } S}$ if and only if $x = x(S)$. Always $x(S) \in \overline{\text{sp } S}$ and $(x - x(s)) \perp \overline{\text{sp } S}$.

Proof: We have $x(S) = \sum_{i=1}^{\infty} (x, u_i)u_i$ for some countable subset of S; this sum converges by the Riesz-Fischer theorem if and only if $\sum_{i=1}^{\infty} |(x, u_i)|^2 \le +\infty$. However, by Bessel's inequality, $\sum_{i=1}^{\infty} |(x, u_i)|^2 \le \|x\|^2$, i.e., $x(S)$ always exists. Clearly $x(S) \in \overline{\text{sp } S}$. Let $s \in S$. Then

$$\left(x - x(S), s\right) = (x,s) - \sum_{u \in s} (x,u)(u,s) = 0$$

so $\left(x - x(S)\right) \perp S$ hence $\left(x - x(S)\right) \perp \overline{\text{sp} S}$. If $x \in \overline{\text{sp} S}$ then $x - x(S) \in \overline{\text{sp} S}$ and so $\|x - x(S)\|^2 = 0$ by what was just shown, i.e. $x = x(S)$.

The next idea is extremely important.

II.10 Definition.   An orthonormal set S in X is *complete* if $S \subset T$ and T is another orthonormal set in X then $S = T$.

That is, a complete orthonormal set is a set which is maximal with respect to being orthonormal. This type of completeness has *nothing* to do with the completeness of the underlying space.

We have tried hard to keep these lectures self-contained. We cheated a bit with the notion of countable sets and now we blatantly pull a result we need from logic. We only use this result once but the consequence is of crucial importance to the remainder of the lectures.

Let P be a set and R a relation on P satisfying for $x, y, z \in P$

(reflexive) xRx

(antisymetric) xRy, yRx implies $x = y$

(transitive) xRy, yRz implies xRz.

Then (P, R) is called a *partially ordered* set. The set P is *linearly ordered* if for $x, y \in P$ either xRy or yRx. If $S \subset P$ then $m \in P$ is an *upper bound for S* if sRm for all $s \in S$. An element $m \in P$ is *maximal* provided $a \in P$ and mRa implies $m = a$.

We now come to one of the most used results in all mathematics. Of course, we state the result without proof (curiously, it was stated without proof by the man whose name it carries!).

II.11 Zorn's Lemma. Let P be a partially ordered set and suppose every linearly ordered subset S has an upper bound in P. Then P has at least one maximal element.

Here is our fundamental application.

II.12 Theorem. Let $X \neq (0)$ be an inner-product space. Then X contains a complete orthonormal set.

Proof: Let $x \neq 0$ be in X. Then $S = \left\{ \frac{x}{\|x\|} \right\}$ is an orthonormal set. Let P be the collection of all orthonormal sets containing S and ordered by inclusion. Let P_0 be a linearly ordered subset of P and consider $S_0 = \bigcup_{U \in P_0} U$. Let $x, y \in S_0$. Since P_0 is linearly ordered x, y are in the same U for some U in P_0. Thus $x \perp y$ and so S_0 is an orthonormal set which is clearly an upper bound for P_0. By Zorn's lemma P has a maximal element T. Clearly (why?) T is a complete orthonormal set in X.

Next we see what the completeness of an orthonormal set S means in terms of sp S.

II.13 Theorem. Let S be orthonormal in X. If $\overline{\text{sp}S} = X$ then S is complete. If H is a Hilbert space and S is a complete orthonormal set then $\overline{\text{sp}S} = H$.

Proof: First suppose S is not complete. Then there is a $y \in X \backslash S$ with $\|y\| = 1$ and $y \perp S$. Clearly $y \notin \overline{\text{sp}S}$ so $X \neq \overline{\text{sp}S}$.

Now suppose H is a Hilbert space and S is a complete orthonormal set in H. Let $x \in H \backslash \overline{\mathrm{sp}S}$ and let $x(S)$ have the meaning of II.9. Then $(x - x(S)) \perp \overline{\mathrm{sp}S}$ and $(x - x(S)) \notin \overline{\mathrm{sp}S}$ (since $x(S)$ is in $\overline{\mathrm{sp}S}$); in particular $x - x(s) \neq 0$ so $S \cup \left\{ \frac{x - x(S)}{\|x - x(S)\|} \right\}$ is an orthonormal set enlarging S so S is not a complete orthonormal set. This contradiction proves the result.

We've looked at famous inequalities so now it's time for a famous equality.

II.14 Parseval's equality.   Consider the statement

$$(*) \qquad\qquad \|x\|^2 = \sum_{u \in S} |(x, u)|^2$$

for each $x \in H$ and some orthonormal set S in H. Then $(*)$ holds if and only if S is a complete orthonormal set.

Proof: Suppose $(*)$ holds and S is not complete. Then there is an x, $\|x\| = 1$, $x \perp S$; so, $1 = \|x\|^2 = \sum_{u \in S} |(x, u)|^2 = 0$. Thus if $(*)$ holds S is complete.

If S is complete, $x = x(S)$ by II.9 and so

$$\|x\|^2 = \left(\sum_{u \in S}(x, u)u, \sum_{v \in S}(x, v)v \right) = \sum_{u, v \in S}(x, u)(\overline{x, v})(u, v) = \sum_{u \in S}|(x, u)|^2.$$

Putting the Parseval equality together with Bessel, Riesz-Fischer, etc., we obtain (essentially) the structure of Hilbert spaces.

II.15 (Summary).   Let H be a Hilbert space. Then

(1) H possesses a complete orthonormal set S;

(2)   for $x \in H$, $x = \sum_{u \in S}(x, u)u$, where the convergence is unconditional, and

$$\sum_{u \in S}|(x, u)|^2 < +\infty :$$

(3)   if H is separable any complete orthonormal S is countable, say $S = \{u_n\}$ and $x = \sum_{n=1}^{\infty}(x, u_n)u_n$, the series an "ordinary" series with the property that all re-arrangements of this series converge to x;

(4) in all cases, if S is complete orthonormal set in H

$$\|x\|^2 = \sum_{u \in S} |(x, u)|^2.$$

We have considered the orthonormal set (e_γ) in $\ell_2(\Gamma)$. The student should check that (e_γ) is, in fact, a complete orthonormal set $\ell_2(\Gamma)$. In particular for $a = (a_n)$ in ℓ_2

$$a = \sum_{n=1}^{\infty} a_n e_n.$$

We will call (e_n) the *canonical* complete orthonormal set for ℓ_2.

Before leaving this material we should look at an important pre-Hilbert space.

II.16 Example. Consider $C[a, b]$, the space of complex-valued continuous functions on $[a, b]$ but with norm given by

$$\|f\|_2 = \left[\int_a^b |f(t)|^2 dt \right]^{1/2}.$$

The inner-product is (obviously)

$$(f, g) = \int_a^b f(t) \overline{g(t)} \, dt.$$

Then $(C[a, b], \| \bullet \|_2)$ is an inner-product space but is not complete. (The student should certainly check this!). To find a complete version of this space (whatever that may mean) requires Lebesgue measure theory. Thus we leave it incomplete. In the space $(C[0, 2\pi], \| \bullet \|_2)$ the functions $\frac{1}{\sqrt{2\pi}} e^{int}$ ($n = 0, \pm 1, \pm 2, \ldots$) are an orthonormal set. Recall that $e^{int} = \cos nt + i \sin nt$ and that integration of e^{int} is accomplished by

$$\int_0^{2\pi} e^{int} dt = \int_0^{2\pi} \cos nt \, dt + i \int_0^{2\pi} \sin nt \, dt.$$

In particular $\frac{1}{2\pi} \int_0^{2\pi} e^{int} e^{-int} dt = \delta_{mn}$. We will have more to say about this type of integration in Chapter V.

Remarks, Exercises and Hints

The notion of orthogonality goes back to vector analysis and Euclidean geometry. From the point of view of functions, d'Alembert (1747), Euler (1748), and D. Bernoulli (1753) gave various solutions to the so- called vibrating string problem:

$$\frac{\partial^2 u}{\partial t^2} = \frac{b^2 \partial^2 u}{\partial x^2},$$

b a positive constant. Bernoulli's solution took the form

$$u = b_1 \sin x \cos bt + b_2 \sin 2x \cos 2bt + \cdots.$$

In 1807 Fourier announced that any function $f(x)$ could be represented

$$(*) \qquad f(x) = \frac{1}{2\pi} \int_{-\pi}^{\pi} f(x)dx + \sum_{n=1}^{\infty} a_n \cos nx + b_n \sin nx$$

where

$$a_n = \frac{1}{\pi} \int_{-\pi}^{\pi} f(x) \cos nx\, dx, \, b_n = \frac{1}{\pi} \int_{-\pi}^{\pi} f(x) \sin nx\, dx.$$

While this statement is not true (nor quite what Fourier said), because of his work we now call a series of the form $(*)$ a Fourier series. The study of Fourier series led to the general theory of orthogonal functions and to an area of Mathematics called Harmonic Analysis (remember the vibrating string).

1. Show that

$$\int_{-\pi}^{\pi} \sin mx \sin nx\, dx = 0 \text{ if } m \neq n$$

$$\int_{-\pi}^{\pi} \cos mx \cos nx\, dx = 0 \text{ if } m \neq n$$

$$\int_{-\pi}^{\pi} \sin nx\, dx = 0, \int_{-\pi}^{\pi} \cos nx\, dx = 0$$

$$\int_{-\pi}^{\pi} \sin^2 nx\, dx = \pi, \int_{-\pi}^{\pi} \cos^2 nx\, dx = \pi \text{ and finally}$$

$$\int_{-\pi}^{\pi} \sin nx \cos nx\, dx = 0.$$

Interpret these relations in $(C[-\pi, \pi], || \bullet ||_2)$.

2. Find the Fourier series of the function

$$f(x) = \frac{x^2}{4} \text{ where } -\pi \leq x \leq \pi.$$

[HINT] Since f is even i.e. $f(x) = f(-x), b_n = 0$. The problem reduces to finding a_n.

3. Using exercise 2, deduce a famous result of Euler:

$$\frac{\pi^2}{6} = \frac{1}{1^2} + \frac{1}{2^2} + \frac{1}{3^2} + \frac{1}{4^2} + \cdots\cdots\cdots.$$

(The series found in (2) does in fact converge to $f(x)$.)

4. Prove Bessels' inequality in terms of Fourier coefficients for $f \in C[-\pi, \pi]$.

$$\frac{1}{2}\left[\frac{1}{\pi}\int_{-}^{\pi} \pi \, f(x)dx\right]^2 + \sum_{n=1}^{\infty}(a_n^2 + b_n^2) \leq \frac{1}{\pi}\int_{-\pi}^{\pi}[f(x)]^2 dx$$

5. If $f \in C[-\pi, \pi]$ has Fourier coefficients $(a_n)(b_n)$ show that

$$\lim_{n\to\infty} a_n = \lim_{n\to\infty} b_n = 0.$$

Continuity is not really needed in the above exercises, but general conditions on functions or their Fourier coefficients takes us to far afield.

Now to orthogonality in abstract Hilbert space. Let H be a Hilbert space.

6. In problem 5 of the first exercise set you showed that if $x, y \in H$ real and $||x|| = ||y||$ then $(x + y, x - y) = 0$. Is this true if H is complex?

7. If $x, y \in H$ and $x \perp y$ show that the Pythagorean theorem holds:

$$||x + y||^2 = ||x||^2 + ||y||^2.$$

If H is real show that the converse is true. Is the converse true if H is complex?

8. If $x, y \in H$ then $x \perp y$ if and only if $||x + \alpha y|| \geq ||x||$ for every scalar α.

9. If $x, y \in H$ then $x \perp y$ if and only if $||x + By|| = ||x - By||$ for every scalar B.

10. (Minimal property). Let $x_1 \cdots x_n$ be orthonormal in H. Let $x \in H$ and $y = \alpha_1 x_1 + \cdots + \alpha_n x_n$ where the $\alpha'_i s$ are scalars. Show that $||x - y||$ is minimum if and only if

$$\alpha_i = (x, x_i); \; i = 1, \cdots n.$$

Recall that a set S in vector space is finitely linearly independent provided any finite subset F of S is a linearly independent set. We end these exercises with two important results in Hilbert space theory.

11. (Gram-Schmidt Process) Suppose that either

   (i) $\{x_1 \cdots x_n\}$ is a finite linearly independent set in H or

   (ii) x_n is a countable finitely linearly independent set in H.

Then in case (i) there is an orthornormal set $\{y_1 \cdots y_n\}$ in H with

$$\mathrm{sp}\{x_i : i = 1, \cdots n\} = \mathrm{sp}\{y_i : i = 1, \cdots n\}.$$

In case (ii) there is a countable orthonormal set (y_n) in H with

$$\mathrm{sp}\{x_i : i = 1, 2, \cdots\} = \mathrm{sp}\{y_i : i = 1, 1, \cdots \cdots\}.$$

[HINT] None of the x_i are zero. Let $y_1 = \frac{x_1}{||x_1||}$. Define the y_i recursively by the following: If

$$z_{n+1} = x_{n+1} - \sum_{i=1}^{n} (x_{n+1}, y_i) y_i.$$

Then

$$y_{n+1} = \frac{z_{n+1}}{||z_{n+1}||}.$$

Here is an important application of the Gram-Schmidt process.

12. Prove that if H is an infinite dimensional separable Hilbert space then H contains a countable complete orthonormal set.

[HINT] Since H is separable, H contains a countable dense $\{x_n\}$. Let z_1 be the first non-zero element in $\{x_n\}$. Let z_2 be the first x_n not in sp $\{z_1\}$ and let z_{k+1} be the first x_n not in sp $\{z_1, \cdots z_k\}$. Gram-Schmidt the set $\{z_n\}$.

13. For convenience we assumed H was a Hilbert space. Completeness (of H) is not necessary in any of these exercises.

III. ISOMORPHISMS AND ISOMETRIES

Recall that if X and Y are linear spaces then $T : X \to Y$ (this symbolism means that T is a function with domain X and range in Y) is a *linear transformation* (or *linear operator*) if for x, $y \in X$, α, $\beta \in \mathbb{C}$, $T(\alpha x + \beta y) = \alpha T(x) + \beta T(y)$. If X and Y are normed linear spaces $T : X \to Y$ is an *isomorphism* (more precisely, a linear topological isomorphism) if T is one-to-one, onto and T and T^{-1} are continuous. Moreover, if T is an isomorphism and $\|Tx\| = \|x\|$ for each $x \in X$ we say T is an *isometry*.

Let us now write $\ell_2(n)$ for n-dimensional Hilbert space, i.e. $\mathbb{C}_n$ with the norm of $a = (a_1, \ldots, a_n)$ given by $\|a\| = \left(\sum_{i=1}^{n} |a_i|^2 \right)^{1/2}$ (or $\mathbb{R}_n$ with the same norm - real Hilbert space!).

Suppose X is an inner-product space of finite dimension n. By Parseval's equality $\|x\| = \sum_{i=1}^{n} |(x, u_i)|^2$ for any complete orthonormal set $\{u_1, \ldots, u_n\}$. The student should check that our earlier arguments show that a complete orthonormal set must be a basis for X and so in this case must have n elements. Define

$$T : X \to \ell_2(n) \quad \text{by} \quad Tx = \big((x, u_i)\big).$$

Parseval equality merely says that T is an isometry. Thus the linear structure and the norm structure are completely preserved. This is, for each positive integer n, usual Euclidean space $\mathbb{C}_n$ (or $\mathbb{R}_n$ in the real case) is the "only" Hilbert space of dimension n.

Now let H be a separable infinite dimensional Hilbert space and let S be a complete orthonormal set in H. Then S must be countable, say $S = \{u_i\}$. Then, again by Parseval's equality,

$$\|x\|^2 = \sum_{i=1}^{\infty} |(x, u_i)|^2$$

and defining

$$T : H \to \ell_2 \quad \text{by} \quad Tx = \big((x, u_i)\big)$$

we see that T is onto by the Riesz-Fischer theorem and, again by Parseval's equality we see that T is an isometry. Thus, in this sense, ℓ_2 is the "only" separable Hilbert space. What did we actually show? All separable infinite dimensional Hilbert spaces (over the complex numbers) are isometric (more precisely, isometrically isomorphic) to the complex sequence space ℓ_2. Of course, a similar statement holds for real spaces.

We should, in all honesty, point out that other realizations of (separable) Hilbert spaces are also very important, especially from the point of view of generating linear operators.

However, we are interested in *representing* a certain class of operators and, from this point of view, ℓ_2 will be enough.

We remark that if one knows a bit about the arithmetic of cardinal numbers one can show that if H is a Hilbert space then *all* complete orthonormal sets in H have the same cardinal number and thus the same argument as above shows that H is isometric to $\ell_2(\Gamma)$ where the cardinal number of Γ is that of a complete orthonormal set for H. Thus, in the sense of isometry, for each cardinal number α there is "one" Hilbert space: namely $\ell_2(\Gamma)$ where the cardinality of Γ is α. This fact is not used in these notes and thus need not be understood.

Remarks, Exercises and Hints

It is fun to look back at characterizations of Hilbert Space since the 1935 result of Jordan and von Neumann showing that H is an inner product space if and only if the parallelogram law: $||x + y||^2 + ||x - y||^2 = 2[||x||^2 + ||y||^2]$ holds. Many researchers used results known to be true for plane and three-dimensional Euclidean geometry. The idea is to place properties on the norm in a normed linear space that arise from geometrical considerations. We give some exercises gleaned from these gems of the past. The proof of the necessity in almost every case consisted of showing that the stated condition on the norm implied the parallelogram law. Some of the proofs of necessity were quite complicated. Yet, all the conditions below characterize real inner-product spaces. We ask only that you show that real inner product spaces have the indicated property. Thus, let H be real inner product space.

1. (Ficken 1944): If $||x|| = ||y||$ then $||ax + by|| = ||bx + ay||$ for all a, b.

   As an aside, Ficken gives a laborious induction argument that in an inner product space $||x|| = ||y|| = \frac{1}{2}||x + y||$ implies $x = y$.

2. Show that this last remark follows immediately from the parallelogram law. Show that the remark is false if the space is not an inner product space, e.g. $\mathbb{R}_n$ with the norm $||x||_1$.

3. (James 1947) If $||x|| = ||y||$ then
   $$\lim_{n \to \infty} \{||nx + y|| - ||x + ny||\} = 0.$$

4. (Lorch 1948) There is a constant $c \neq 0, 1, -1$ such that if
   $$||x + y|| = ||x - y|| \text{ then } ||x + cy|| = ||x - cy||.$$

5. (Lorch 1948) (Isosceles Triangle Theorem) If
   $$x, y, z \in H, x + y + z = 0$$

and
$$||x|| = ||y||$$

then
$$||x - z|| = ||y - z||.$$

IV. BOUNDED LINEAR OPERATORS ON HILBERT SPACE

Let H be a Hilbert space and $T : H \to H$ a linear transformation, from now on called a *linear operator* . The operator T is *bounded* if there is an $M > 0$ such that $\|Tx\| \leq M\|x\|$ for each $x \in H$. Let

$$\mathfrak{L}(H) = \{T : H \to H \mid T \text{ is linear and bounded}\}.$$

(This set is also sometimes denoted by $B(H)$ in the literature.) Let $\|T\| = \sup\{\|Tx\| \mid \|x\| \leq 1\}$. We give some exercises which should be worked immediately.

Exercise 1. Let $T : H \to H$ be linear. The following are equivalent:
    (a) T is bounded;
    (b) T is continuous at 0; and,
    (c) T is continuous on all of H.

Exercise 2. If $T \in \mathfrak{L}(H)$, $\|T\|$ is a norm and $\mathfrak{L}(H)$ with this norm is a Banach space (that is, it is complete).

Exercise 3. If $T \in \mathfrak{L}(H)$ then $\|Tx\| \leq \|T\|\|x\|$ for all $x \in H$ and so, in particular, T is *uniformly continuous* on H.

Suppose $A \in \mathfrak{L}(H)$. A is *invertible* if there is a $B \in \mathfrak{L}(H)$ such that $AB = BA = I$, where I is the identity operator on H. The operator B is called the *inverse* of A and we write $B = A^{-1}$.

Exercise 4. If $H = \ell_2(n)$ for some integer n then $\mathfrak{L}(H)$ can be identified with the set of all $n \times n$ complex-valued matrices. The invertible operators correspond to the non-singular matrices. (See Chapter X.)

Exercise 5. If A, B, $C \in \mathfrak{L}(H)$ and $AB = CA = I$ then $B = C = A^{-1}$.

Exercise 6. If A, $B \in \mathfrak{L}(H)$ are invertible then $(AB)^{-1} = B^{-1}A^{-1}$.

Exercise 7. If $A \in \mathfrak{L}(H)$ is invertible and $A^n x = A\big(A^{n-1}(x)\big)$, $n \geq 2$, then A^n is invertible for all positive integers n and $(A^n)^{-1} = (A^{-1})^n$.

We adopt the following notation: If $A \in \mathfrak{L}(H)$, $\mathcal{R}(A)$ denotes the range of A, i.e., $\mathcal{R}(A) = \{Ax \mid x \in H\}$.

IV.1 Lemma.   Suppose $A \in \mathfrak{L}(H)$ and there is an $\alpha > 0$ such that $\|Ax\| \geq \alpha\|x\|$ for all $x \in H$. Then $\mathcal{R}(A)$ is closed.

Proof: Let $y \in \overline{\mathcal{R}(A)}$ and choose $(x_n) \in H$ such that (Ax_n) converges to y. In particular (Ax_n) is Cauchy and so for $\varepsilon > 0$ there is an N such that $m, n \geq N$ implies

$$\|Ax_n - Ax_m\| = \|A(x_n - x_m)\| < \varepsilon.$$

By Hypothesis, $\|x_n - x_m\| < \frac{\varepsilon}{\alpha}$ and so (x_n) is Cauchy and, since H is complete, there is an $x \in H$ such that $\lim_{n \to \infty} x_n = x$. By continuity, $\lim_{n \to \infty} Ax_n = Ax$. Thus $Ax = y$ and so $\mathcal{R}(A)$ is closed.

We can now characterize invertible operators. These operators play an important role (understatement!) in the spectral theory to follow.

IV.2 Theorem. *Let $T \in \mathfrak{L}(H)$. Then, T is invertible if and only if $\mathcal{R}(T) = H$ and there is $\alpha > 0$ such that $\|T\| \geq \alpha\|x\|$ for all $x \in H$.*

Proof: If T is invertible and $y \in H$ let

$$x = T^{-1}y \quad (\text{don't forget:} \quad T^{-1} \in \mathfrak{L}(H)!).$$

Thus $Tx = y$ and so $\mathcal{R}(T) = H$. Also, for $x \in H$, $\|x\| = \|T^{-1}Tx\| \leq \|T^{-1}\|\|Tx\|$ so with $\alpha = \frac{1}{\|T^{-1}\|}$ we have $\|Tx\| \geq \alpha\|x\|$. Now suppose $\mathcal{R}(T) = H$ and $\|Tx\| \geq \alpha\|x\|$ for some $\alpha > 0$ and all $x \in H$. Clearly T is one-to-one (why?) and, by hypothesis, onto. Define $S : H \to H$ by $Sy = x$ where $Tx = y$. Then S is well defined and linear (why?). Also $\|TS(y)\| \geq \alpha\|Sy\|$ so $\|Sy\| \leq \frac{1}{\alpha}\|y\|$. Hence $S \in \mathfrak{L}(H)$. By definition $S = T^{-1}$.

We look now at an important class of non-invertible operators. Let $P \in \mathfrak{L}(H)$ be such that $P^2 = P$ (i.e. $P(Px) = P(x)$), or, P is the identity on its range) then P is called a *projection*. From $\|Px\| = \|P(P(x))\| \leq \|P\|\|P(x)\|$ we see that for a projection P, $\|P\| \geq 1$.

Let H_0 be a *closed* subspace of the Hilbert space H. It is easy to see (check this!) that H_0 also forms a Hilbert space with the inner-product norm inherited from H. Thus there is a complete orthonormal set S in H_0 (complete for H_0!) by II.12. This orthonormal set S may be extended to a complete orthonormal set in H (by the proof of II.12). Thus let $T \subset H \backslash H_0$ be such that $S \cup T$ is a complete orthonormal set in H. Now $H_0 = \overline{\text{sp}S}$. Let $N = \overline{\text{sp}\,T}$. If $x \in H_0$, $y \in N$, $x \perp y$ and hence $H_0 \perp N$ (which means each element of H_0 is orthogonal to each element of N). We call N the *orthogonal complement* of H and write $N = H_0^{\perp}$. What is the meaning of all this? Let $x(S)$, $x(T)$ have the meaning of I.9, i.e., $x(S) = \sum_{u \in S} (x, u)u$ and $x(T) = \sum_{u \in T} (x, u)u$. Clearly from our previous results, $x = x(S) + x(T)$. We define $P_S(x) = x(S)$ and $P_T(x) = x(T)$. Thus $P_S(x) = \sum_{u \in S} (x, u)u$ and so

$$P_S(P_S(x)) = \sum_{v \in S} \left(\sum_{u \in S} (x, u)u, v \right) v = \sum_{u \in S} (x, u)u = P_S(x)$$

and similarly for P_T, i.e., P_S and P_T are projections. We call them *orthogonal projections* for obvious reasons. Also

$$\|P_S(x)\|^2 = \left\| \sum_{u \in S} (x, u)u \right\|^2 = \sum_{u \in S} |(x, u)|^2 \quad \text{(Parseval)} \leq \sum_{u \in S} |(x, u)|^2$$
$$+ \sum_{v \in T} |(x, v)|^2 = \|x\|^2 \quad \text{(Parseval)}$$

i.e., $\|P_S\| \leq 1$. Similarly, $\|P_T\| \leq 1$. Since any bounded projection P has $\|P\| \geq 1$, we have that for these *orthogonal projections*, P_S, P_T, $\|P_S\| = 1$, $\|P_T\| = 1$.

We have proved the following result.

IV.3 Theorem. *On each closed subspace H_0 of a Hilbert space H we can find an orthogonal projection in $\mathfrak{L}(H)$ of norm 1. Moreover $H = H_0 \oplus H_0^{\perp}$ where $H_0^{\perp} = \{x | P(x) = 0\}$ where $P : H \to H_0$ is given by P_S above. (Here $H = H_0 \oplus H_0^{\perp}$ means every $x \in H$ can be written uniquely as $x = a + b$, $a \in H_0$, $b \in H_0^{\perp}$ and that the projection operations $x \to a$ and $x \to b$ are continuous.)*

Curiously Hilbert space is, isomorphically, the only space with this property. Indeed, Lindenstrauss and Tzafriri have shown that if Y is a Banach space and onto every closed subspace of Y there is a bounded linear projection, then Y is isomorphic to a Hilbert space.

For a Banach space Y let Y^* denote the linear space of all continuous linear operators from Y to the scalar field $\mathbb{C}$ (hereafter, such objects are called *linear functionals*) and for $f \in Y^*$ let

$$\|f\| = \sup\{|f(x)| : \|x\| \leq 1\}.$$

Exercise 8. The space Y^* equipped with the above norm is a Banach space. (Actually Y need not be complete for this statement to be true.)

The space Y^* is the *dual* (or *conjugate*) space of Y. Let Y be an inner product space.

Exercise 9. Let $f \in Y^*$, $f \neq 0$. Prove there is a $z_0 \in Y$ such that for all $x \in Y$ there is a unique scalar α and $y \in \ker f = \{y \in Y | f(y) = 0\}$ such that $x = \alpha z_0 + y$, $f(x) = \alpha$, and $z_o \in (\ker f)^{\perp}$.

[HINT] Choose $z \in (\ker f)^{\perp}$ with $f(z) \neq 0$. Let $z_0 = \frac{z}{f(z)}$ and if $f(x) = \alpha$ write $x = \alpha z_0 + (x - \alpha z_0)$.

Characterizing the *conjugate (dual)* space Y^* is an important task in Banach space theory. What is the dual of a Hilbert space H?

IV.4 Riesz Representation Theorem. *Let H be a Hilbert space.*
    (a) Let $f \in H^*$. Then there is a unique $y \in H$ such that $f(x) = (x, y)$
        for all $x \in H$. Moreover, $\|f\| \leq \|y\|$.
    (b) Let $y \in H$. Define $f_y(x) = (x, y)$. Then $f_y \in H^*$ and $\|f_y\| \leq \|y\|$.

Proof: (a) Let $f \in H^*$, $f \neq 0$. By exercise 9, there is a $z, \|z\| = 1$, in $(\ker f)^{\perp}$ such that for $x \in H$ there is a unique scalar α and $\omega \in \ker f$ with $x = \alpha z + \omega$.

Let $y = \overline{f(z)}z$. Then

$$(x, y) = (\omega + \alpha z, y) = \alpha f(z).$$

Also $f(x) = \alpha f(z) + f(\omega) = \alpha f(z)$. To show uniqueness, suppose $f(x) = (x, y_1) = (x, y_2)$ for all $x \in H$. Then $(x, y_1 - y_2) = 0$ for all $x \in H$, in particular for $x = y_1 - y_2$, so $\|y_1 - y_2\|^2 = 0$, i.e., $y_1 = y_2$. Also, $|f(x)| = |(x, y)| \leq \|x\|\|y\|$ by the Cauchy-Schwarz-Bunyakovsky inequality so $\|f\| \leq \|y\|$.

(b) If, for $y \in H$, $f_y(x) = (x, y)$ then clearly $f_y \in H^*$ (check the properties of $(\bullet, \bullet)$!) and $|f_y(x)| = |(x, y)| \leq \|x\|\|y\|$ as above, i.e., $\|f_y\| \leq \|y\|$.

Exercise 10. Define $\phi : H \to H^*$ by $\phi(y) = f_y$. Is ϕ an isomorphism? Examine this carefully!

Exercise 11. Let M be a closed proper subspace of H. If $y \in H \backslash M$ show that there is a $z \in H$ such that $(y, z) \neq 0$ and $(m, z) = 0$ for all $m \in M$.

[HINT] If $(y, z) = 0$ for all $z \in M^\perp$ then from $H = M \oplus M^\perp$ it follows that $y = 0$.

We now can define the important concept of adjoint of a continuous linear operator on Hilbert space.

IV.5 Definition. Let $T \in \mathcal{L}(H)$. Define for $y \in H$, $T^*y \in H^*$ as follows:

$$(T^*y)(x) = (Tx, y).$$

By the Riesz representation theorem there is a unique $z \in H$ with $(T^*y)(x) = (Tx, y) = (x, z)$. We write $z = T^*y$ under this identification (so now $T^*y \in H$!) and thus $(Tx, y) = (x, T^*y)$. The operator T^* is called the adjoint of T.

The student should check what T^* means if T is an $n \times n$ matrix (viewed as a continuous linear operator) on $\ell_2(n)$.

The adjoint operator plays a very important role in all that follows.

Remarks, Exercises, and Hints

Kakatani (in 1939) proved that a normed linear space of dimension at least three is an inner product space if and only if every two dimensional subspace is the range of a norm one projection. We showed that every closed supspace of a Hilbert space is the range of a norm one projection. Thus the Kakatani result implies the fact that if the space has dimension at least three and that if there is a norm one projection onto every two dimensional subspace, then there is a norm one projection onto *every* closed subspace.

Our proof of the fact that there is a norm one projection onto each closed subspace of a Hilbert space H, uses the existence of an orthogonal expansion for each $x \in H$. The standard proof of this norm one projection result takes a different approach.

Let H be a Hilbert space.

1. A set K in H (or any linear space for that matter) is convex if $tx + (1-t)y \in K$ for all $x, y \in K$ and all t, $0 \le t \le 1$.

   Show that if K is a closed non-empty convex set in H then for $x \in H$ there is a unique $y \in K$ with $||x - y|| = \text{dist}(x, K) = \inf\{||x - k|| : k \in K\}$.

   [HINT] Let $d = \text{dist}(x, K)$. Choose $k_n \in K$ such that $\lim_{n \to \infty} ||x - k_n|| = d$. Show that $\{k_n\}$ is a convergent sequence by applying the parallelogram identity:

   $$||k_n - k_m||^2 = 2\left[||x - k_m||^2 + ||x - k_n||^2 - 2||x - \frac{1}{2}(k_m + k_n)||^2\right].$$

   Now use the fact that $\frac{1}{2}(k_m + k_n) \in K$ and the definition of d. Let $y = \lim_{n \to \infty} k_n$. Clearly $||x - y|| = d$. For uniqueness again use the parallelogram identity and the convexity of K.

2. Let Y be a closed subspace of H. For $h \in H$ let y_0 be the unique element of Y satisfying $d = \text{dist}(h, Y) = ||h - y_0||$. Show that $h - y_0 \in Y^\perp$.
   [HINT] For $y \in Y$ and $\alpha \in \mathbb{C}$.

Compute:

$$0 \le ||h - (y_0 + \alpha y)||^2 - ||h - y_0||^2$$
$$= -\alpha(y, h - y_0) - \bar{\alpha}(h - y_0, y) + |\alpha|^2 ||\alpha y||^2.$$

Let r be a real number and $\alpha = r(h - y_0, y)$ and deduce from the above computation that

$$h - y_0 \in Y^{\perp}.$$

3. Show that Exercise 2 implies that if Y is a closed subspace of H then $H = Y \oplus Y^{\perp}$ with the meaning of $\oplus$ as explained in the text.

4. Let Y be a subspace of H. Show that Y is closed if and only if $(Y^{\perp})^{\perp} = Y$.

5. In the proof of the Riesz representation theorem (IV.4) show that if $f \in H^*, f \neq 0, (\ker f)^{\perp}$ is one-dimensional, i.e. $(\ker f)^{\perp} = \text{sp}\{z\}$ for some $z \in H$.

6. Show that if $f \in (\ell_2)^*$ there is a unique sequence

$$(\beta_i) \in \ell_2$$

such that for

$$\alpha = (\alpha_i) \in \ell_2, f(\alpha) = \sum_{i=1}^{\infty} \alpha_i \beta_i$$

and

$$\|f\| = \left(\sum_{n=1}^{\infty} |\beta_n|^2 \right)^{\frac{1}{2}}.$$

[HINT] Let (e_n) be the canonical complete orthonormal set for ℓ_2 and let $\beta_n = f(e_n)$. To see that $(\beta_n) \in \ell_2$, fix n and let

$$\alpha_k = \begin{cases} |\beta_k| \, \text{sgn} \, \bar{\beta}_k & 1 \leq k \leq n \\ 0 & k > n \end{cases}.$$

Recall that $\text{sgn} \, z = \frac{\bar{z}}{|z|}$ if $z \neq 0$, 1 if $z = 0$. Then $\beta_k \alpha_k = |\beta_k|^2 = |\alpha_k|^2$ if $1 \leq k \leq n$. If

$$\alpha = (\alpha_k), \|\alpha\| = \left(\sum_{i=1}^{n} |\beta_k|^2 \right)^{\frac{1}{2}}$$

and

$$f(\alpha) = \sum_{i=1}^{n} \alpha_k \beta_k = \sum_{i=1}^{n} |\beta_k|^2.$$

Thus it follows that $\left(\sum_{i=1}^{n} |\beta_n|^2 \right)^{\frac{1}{2}} \leq ||f||$, i.e. $(\beta_n) \in \ell_2$. Clearly, if $\gamma = (\gamma_n) \in \ell_2$ and $f(\alpha) = \sum_{n=1}^{\infty} \alpha_n \gamma_n$ where $\alpha = (\alpha_n) \in \ell_2$ then $f \in (\ell_2)^*$. Use the Cauchy-Schwarz-Bunyakovsky inequality to conclude that $|f(\alpha)| \leq ||\alpha|| ||\gamma||$. Now put the pieces together.

7. Let $x, y \in H$ correspond to $f, g \in H^*$ via the Riesz representation theorem.

i.e.

$$f(z) = (z, x), g(z) = (z, y) \text{ for all } z \in H.$$

Define $(f, g) = (y, x)$. Show that (f, g) is an inner-product on H^* and show that H^* is isometrically isomorphic to H.

V. ELEMENTARY SPECTRAL THEORY

General spectral theory (whatever it is) can be rather difficult. In these lectures we will deal almost exclusively with eigenvalues, thus simplifying things considerably.

V.1 Definition. Let $T \in \mathcal{L}(H)$ and $\lambda \in \mathbb{C}$, $\lambda \neq 0$. Then λ is an *eigenvalue* of T provided there is an $x \in H$, $x \neq 0$ such that $Tx = \lambda x$. The vector x is called an *eigenvector associated to* λ. More generally, an $x \in X$, $x \neq 0$, is a generalized eigenvector for $\lambda \neq 0$ if there is a positive integer m with

$$(T - \lambda I)^m (x) = 0.$$

We will avoid some difficulties by assuming we are always working with eigenvalues and eigenvectors. For compact operators, to be presently defined and our true concern in these notes, there will be no problems with these different concepts.

Clearly if λ is an eigenvalue of T then $T - \lambda I$, I the identity operator on H, is not invertible.

V.2 Definition. Let $T \in \mathcal{L}(H)$. The *spectrum* of T, written $\sigma(T)$ is defined as follows: $\sigma(T) = \{\lambda \in \mathbb{C} \mid T - \lambda I$ is not invertible$\}$.

In particular for an $n \times n$ matrix T, $T - \lambda I$ is not invertible if and only if the determinant,

$$\det(T - \lambda I) = 0.$$

Thus, in the finite dimensional case $\sigma(T)$ is just the eigenvalues of T (since $\det(T - \lambda I)$ is an nth degree polynomial whose roots are the eigenvalues of T).

The complement $\rho(T)$ of the spectrum of T, $\rho(T) = \mathbb{C} \backslash \sigma(T)$, is called the *resolvent* of T.

Much of what follows comes from the following easily proved fact. If $|r| < 1$ the series $\sum_{r=0}^{\infty} ar^n$ converges to $\frac{a}{1-r}$. The student should observe that this is just the convergence of a geometric series. In particular if $z \in \mathbb{C}$, $|z| < 1$ then

$$(1 - z)^{-1} = \sum_{n=0}^{\infty} z^n \qquad (z^0 \text{ is defined to be } 1).$$

Let $T \in \mathcal{L}(H)$ and suppose $\|T\| < 1$. Then $\sum\limits_{n=0}^{\infty} T^n$ (T^0 is defined to be I, the identity operator) converges in $\mathcal{L}(H)$ since $\sum\limits_{n=0}^{\infty} \|T\|^n$ is a convergent geometric series and so $\left(\sum\limits_{n=1}^{m} T^n \right)$ is Cauchy in $\mathcal{L}(H)$, hence convergent, since $\mathcal{L}(H)$ is complete.

For the moment, write $S = \sum\limits_{n=0}^{\infty} T^n$ where $\|T\| < 1$. What is S? Let us calculate $(I - T)S$:

$$(I - T)(S) = S - TS = \sum_{n=0}^{\infty} T^n - \sum_{n=0}^{\infty} T^{n+1} = I.$$

Also, $S(I - T) = I$ and so $S = (I - T)^{-1}$. We have derived the C. Neumann expansion of $(I - T)^{-1}$ with $\|T\| < 1$. Since this easy result will be used often we state it more formally.

V.3 C. Neumann expansion. Let $T \in \mathcal{L}(H)$, $\|T\| < 1$. Then $I - T$ is invertible and

$$(I - T)^{-1} = \sum_{n=0}^{\infty} T^n.$$

Moreover

$$\|(I - T)^{-1}\| \le \frac{1}{1 - \|T\|}.$$

We can now prove some interesting results concerning the resolvent of T. For $\zeta \in \rho(T)$ the operator $R_\zeta = (\zeta I - T)^{-1}$ is called the *resolvent operator* (note that R_ζ depends on ζ and T and is only defined for complex numbers in $\rho(T)$).

Let us observe that the Neumann expansion implies that $\rho(T) \ne \emptyset$. Indeed if $\zeta \in \mathbb{C}$ and $|\zeta| > \|T\|$, $\zeta I - T = \zeta(I - \zeta^{-1}T)$ and $\|\zeta^{-1}T\| < 1$ so $(I - \zeta^{-1}T)^{-1}$ exists. Thus $(\zeta I - T)^{-1} = \zeta^{-1}(I - \zeta^{-1}T)^{-1}$.

V.4 Theorem. *Suppose $\zeta \in \rho(T)$ and $|\zeta - \eta| < \|R_\zeta\|^{-1}$. Then $\eta \in \rho(T)$ and*

$$R_\eta - R_\zeta = \sum_{n=1}^{\infty} (\zeta - \eta)^n R_\zeta^n.$$

Moreover,

$$\|R_\eta\| \leq \frac{\|R_\zeta\|}{1 - |\zeta - \eta|\|R_\zeta\|}.$$

Proof: The proof involves a trick which will be used again and again: write $\eta I - T = (\zeta I - T) - (\zeta - \eta)I$. By hypothesis $\|(\zeta - \eta)R_\zeta\| < 1$ and so $[I - (\zeta - \eta)R_\zeta]^{-1}$ exists and is equal to

$$\sum_{n=0}^{\infty} [(\zeta - \eta)R_\zeta]^n.$$

Thus,

$$\begin{aligned}
& \left[R_\zeta[I - (\zeta - \eta)R_\zeta]^{-1}\right](\eta I - T) \\
&= R_\zeta[I - (\zeta - \eta)R_\zeta]^{-1}[(\zeta I - T) - ((\zeta - \eta)I)] \\
&= R_\zeta[I - (\zeta - \eta)R_\zeta]^{-1}[R_\zeta^{-1} - (\zeta - \eta)R_\zeta^{-1}R_\zeta] \\
&= [I - (\zeta - \eta)R_\zeta]^{-1}[I - (\zeta - \eta)R_\zeta] = I.
\end{aligned}$$

Clearly everything commutes (check!) and the assertion is proved. The norm estimate is immediate from V.3.

Exercise 1. Let $T \in \mathcal{L}(H)$ and suppose T^{-1} exist. Prove that if $S \in \mathcal{L}(H)$ and $\|T - S\| < \frac{1}{\|T^{-1}\|}$, then S^{-1} exist and

$$\|S^{-1} - T^{-1}\| \leq \frac{\|T^{-1}\|^2\|T - S\|}{1 - \|T^{-1}\|\|T - S\|}.$$

In particular, on the set of invertible operators in $\mathcal{L}(H)$, the mapping $T \to T^{-1}$ is continuous.
[HINT] Mimic the proof of V.4.

We need yet another topological concept.

A set A in $\mathbb{C}$ is *open* if $\mathbb{C} \backslash A$ is closed. We have shown that $\rho(T)$ is open for any $T \in \mathcal{L}(H)$ (why ?) and thus, that $\sigma(T)$ is closed. We have also shown that $\sigma(T)$ is contained in the disk $\{z \in \mathbb{C} \mid |z| \leq \|T\|\}$, i.e., $\sigma(T)$ is closed and bounded.

V.5 Theorem. *If* $\zeta, \eta \in \rho(T)$

$$R_\zeta - R_\eta = (\eta - \zeta)R_\zeta R_\eta.$$

Proof: Clearly the operators R_ζ and R_η commute (check) and so

$$(R_\zeta - R_\eta)R_\zeta^{-1}R_\eta^{-1} = R_\eta^{-1} - R_\zeta^{-1} = (\eta - \zeta)I.$$

Theorem V.5 implies (together with the norm estimate from V.4) that the function $\phi : \rho(T) \to \mathcal{L}(H)$ given by $\phi(\eta) = R_\eta$ is continuous. This fact is vital to the integration theory which follows and *must* be checked by the student.

Let us list the properties obtained so far (we will need them often):

(i)   If $\zeta \in \rho(T)$ and $|\zeta - \eta| < \|R_\zeta\|^{-1}$ then $\eta \in \rho(T)$;

(ii) If $|\zeta| > \|T\|$ then $\zeta \in \rho(T)$ and $R_\zeta = \zeta^{-1}(I + \zeta^{-1}T + \zeta^{-2}T^2 + \ldots)$;

(iii) If $\zeta, \eta \in \rho(T)$ then $R_\zeta - R_\eta = (\eta - \zeta)R_\zeta R_\eta$; and

(iv) The resolvent mapping $\phi : \rho(T) \to \mathcal{L}(H)$ given by $\phi(\zeta) = R_\zeta$ is continuous.

We now want to define an operator valued integral. A very general theory can be developed mimicking the development of the line integral in complex analysis. For our purposes we need only consider integrals defined on (or along) circles centered at 0 and, briefly, curves consisting of straight lines and circular arcs.

Thus, let C_r be the circle $|z| = r$ in the complex plane $\mathbb{C}$. Clearly $z = re^{i\theta}$, $\theta \in [0, 2\pi]$ (recall that $e^{i\theta} = \cos\theta + i\sin\theta$). Thus $dz = rie^{i\theta}d\theta$ and so $\int_{C_r} f(z)dz$ merely means $r\int_0^{2\pi} f(re^{i\theta})e^{i\theta}id\theta$ and one integrates just as one integrates $\int e^u du$ for real u! In particular if $f(z) = z^n$, $n \geq 0$ then

$$\int_{C_r} f(z)dz = \int_{C_r} z^n dz = \int_0^{2\pi} (re^{i\theta})^n rie^{i\theta}d\theta$$

$$= r^{n+1}\int_0^{2\pi} e^{i(n+1)\theta}id\theta = 0 \qquad n = 0, 1, 2, \ldots .$$

Also for

$$g(z) = z^{-n}, \qquad n = 2, 3, 4, \ldots$$

$$\int_{C_r} g(z)dz = 0. \quad \text{(check!)}$$

However, for $n = 1$

$$\int_{C_r} \frac{dz}{z} = \int_0^{2\pi} \frac{rie^{i\theta}d\theta}{re^{i\theta}} = \int_0^{2\pi} id\theta = 2\pi i.$$

Next observe that if a is inside C_r and $z \in C_r$

$$\frac{1}{z-a} = \frac{1}{z}\left[\frac{1}{1-\frac{a}{z}}\right] = \frac{1}{z}\left[1 + \frac{a}{z} + \frac{a^2}{z^2} + \cdots\right]$$

and the series converges since $|\frac{a}{z}| < 1$. (See the remarks before the Neumann expansion.) We can integrate term by term (refresh your power series!) and by the preceding remarks

$$\int_{C_r} \frac{dz}{z-a} = 2\pi i.$$

If a is outside the circle C_r and $z \in C_r$,

$$\frac{1}{z-a} = -\frac{1}{a}\left[\frac{1}{1-\frac{z}{a}}\right] = -\frac{1}{a}\left[1 + \frac{z}{a} + \left(\frac{z}{a}\right)^2 + \cdots\right]$$

and this series converges since $|\frac{z}{a}| < 1$. Now $\int_{C_r} \frac{dz}{z-a} = 0$.

We seek the operator analog of these results. We define for $T \in \mathcal{L}(H)$ and $C_r = \{z\,|\,|z| = r\}$ *entirely in* $\rho(T)$

$$\int_{C_r} R_\zeta d\zeta = \int_0^{2\pi} \phi\left(re^{i\theta}\right) rie^{i\theta} d\theta$$

where ϕ is the resolvent mapping $\phi(\zeta) = R_\zeta$. The integral, of course, is meant in the sense of partitions: if $0 = \theta_0 < \theta_1 < \cdots < \theta_n = 2\pi$ is a partition P of $[0, 2\pi]$ and $\theta'_j \in [\theta_{j-1}, \theta_j]$, $S(P) = \sum_{j=1}^n \phi\left(re^{i\theta'_j}\right)\left[re^{i\theta_j} - re^{i\theta_{j-1}}\right] \in$ $\mathcal{L}(H)$. By *all* the preceding properties (i)-(iv) plus the completeness of $\mathcal{L}(H)$, if $|P| = \max|\theta_j - \theta_{j-1}|$ then $\lim_{|P|\to 0} S(P)$ exist (this limit in the same sense as for the Riemann integral) and we denote this operator by $\int_{C_r} R_\zeta d\zeta$. Also if $[a, b]$ is a real interval we define $\int_a^b R_t dt$ in the same manner. Clearly $\int_a^b R_t dt = -\int_b^a R_t dt$. To see that $\lim_{|P|\to 0} S(P)$ exists (for C_r or $[a, b]$) it is enough to show that for $\varepsilon > 0$ there is a $\delta > 0$ such that if $|P| < \delta$, $|Q| < \delta$

$$(*) \qquad\qquad \|S(P) - S(Q)\| < \varepsilon$$

(no matter how the intermediate points are chosen). Indeed, using the condition $(*)$ for $\varepsilon = \frac{1}{n}$ and arranging things so that $\delta_n < \delta_m$ if $n > m$, let $|P_n| < \delta_n$. If $n > m$

$$\|S(P_n) - S(P_m)\| < \frac{1}{m} \qquad \text{and so}$$

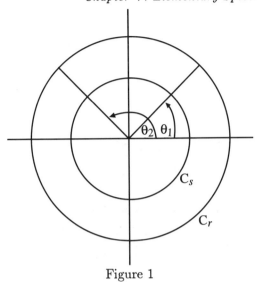

Figure 1

$(S(P_n))$ is Cauchy in $\mathfrak{L}(H)$ and so has a limit S. Again using (∗) it is easy to show that S does not depend on the intermediate choices $\theta'_j \in [\theta_{j-1}, \theta_j]$.

Similarly for $\int_a^b R_\zeta d\zeta$. Let's do two simple calculations. Recall that if $r > \|T\|$, C_r is in $\rho(T)$ and for $\zeta \in C_r$, $R_\zeta = \sum_{n=1}^{\infty} \zeta^{-n} T^{n-1}$. Thus for such r

$$\int_{C_r} R_\zeta d\zeta = \sum_{n=1}^{\infty} T^{n-1} \int_{C_r} \zeta^{-n} d\eta = 2\pi i I,$$

where I is the identity operator.

Now recall also that if $\eta, \zeta \in C_r$, $C_r \subset \rho(T)$ and $|\zeta - \eta| < \|R_\zeta\|^{-1}$,

$$R_\eta = \sum_{n=0}^{\infty} (\zeta - \eta)^n R_\zeta^{n+1} \qquad \text{and so}$$

$$\int_{C_r} R_\eta d\eta = 0. \qquad \text{(why?)}$$

We will do a slightly more difficult calculation which will play an important role in what follows. Consider two concentric circles C_s and C_r with $s < r$ and two fixed angles θ_1, θ_2. (See Figure 1.)

Consider the "wedge" W bounded by C_r, C_s and the rays determined by θ_1, θ_2 so W is given (geometrically) by Figure 2.

Let us integrate $\int_W z^n dz$, $n = 0, 1, 2, \ldots$ (counter-clockwise). Since A is the arc of C_r from θ_1 to θ_2, A has a parametric representation $z = re^{i\theta}$,

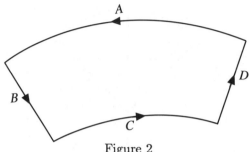

Figure 2

$\theta \in [\theta_1, \theta_2]$. Similarly C has a representation $z = se^{i\theta}$, $\theta \in [\theta_2, \theta_1]$ The ordering here and below is important! Also B is given by $z = xe^{i\theta_2}$, $x \in [r, s]$ and D by $z = xe^{i\theta_1}$, $x \in [s, r]$. Thus $\int_W z^n dz = \int_A z^n dz + \int_B z^n dz + \int_C z^n dz + \int_D z^n dz = 0$.

Exercise 2.   Check that the sum of these four integrals is indeed zero. Just compute using the above representations. For example

$$\int_A z^n dz = \int_{\theta_1}^{\theta_2} [re^{i\theta}]^n rie^{i\theta} d\theta$$

$$= \frac{r^{n+1}}{n+1} \left[e^{i(n+1)\theta_2} - e^{i(n+1)\theta_1} \right]$$

Now suppose the concentric circles and the region between them lie entirely in $\rho(T)$ for some $T \in \mathcal{L}(H)$. (Such a region is called an annular region.)

V.6 Theorem. *For such an annular region A, $\int_{C_r} R_\eta d\eta = \int_{C_s} R_\eta d\eta$.*

Proof: Since A is a compact (i.e. closed and bounded) subset of $\mathbb{C}$ and $\phi(\eta) = \|R_\eta\|$ is continuous on A, $\max\{\|R_\eta\| : \eta \in A\} = M < +\infty$. Thus $\|R_\zeta\|^{-1} \geq \frac{1}{M}$ for all $\zeta \in A$. The idea of the proof is simple but the details are a bit messy: Fix $\zeta_1 \in C_r$. Choose t, $s < t < r$ and θ_1, θ_2 so that ζ_1 is on the segment determined by θ_1 and the wedge W_t^1 between C_r and C_t has the property that if $\eta \in W_t$ then $|\eta - \zeta_1| < \|R_{\zeta_1}\|^{-1}$. (See figure 3.)

Using the above calculation for W and the expansion of R_η, it follows that $\int_{W_t} R_\eta d\eta = 0$. (All our integrals are computed counterclockwise) Now build another wedge to the right of W_t^1, say, W_t^2 with the same properties as W_t^1 with respect to ζ_1. Then $\int_{W_t^2} R_\eta d\eta = 0$.

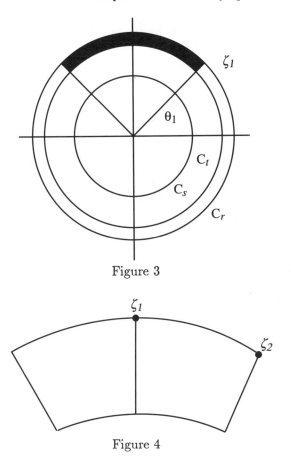

Figure 3

Figure 4

Observe that since we are integrating counterclockwise the direction traversed on the segment containing ζ_1 is "up" for W_t^1 and "down" for W_t^2 and so these integrals cancel one another. Thus the integral over Figure 4 is zero.

We now repeat the process with ζ_2 in place of ζ_1 adding two more wedges. Clearly we can continue this process until we return to W_t^1 at the ray determined by θ_2. Since $\|R_\zeta\|^{-1}$ is bounded away from zero on A, a finite number of steps suffice (check). The integrals on the segments cancel and we obtain (since C_t is traversed clockwise)

$$\int_{C_r} R_\eta d\eta = \int_{C_t} R_\eta d\eta.$$

We can repeat the process with C_t replacing C_r and gradually (in a finite number of steps) "deform" C_r to C_s obtaining the desired result.

We can now prove one of the fundamental results of spectral theory (although we don't really need it!).

V.7 Theorem. *Let $T \in \mathcal{L}(H)$, $H \neq (0)$. Then the spectrum of T is not empty.*

Proof: If $T = 0$, $\sigma(T) = \{0\}$ (why?). Thus let $T \neq 0$ and suppose $\sigma(T)$ is empty. Then $\rho(T) = \mathbb{C}$ and so $C_r \subset \rho(T)$ for all positive r. The above remarks show that $I = \frac{1}{2\pi i} \int_{C_r} R_\zeta d\zeta$ for all such r. (Why?) Letting $r \to 0$ we obtain also $\int_{C_0} R_\zeta d\zeta = 0$, where $C_0 = \{0\}$. So $2\pi i I = 0$ which is nonsense. (One needs (iii) to show that this last limit procedure is valid. The student should check it.)

We have shown that for $T \in \mathcal{L}(H)$, $\sigma(T)$ is a non-empty, closed, bounded $\big(\sigma(T) \subset \{z | |z| \leq \|T\|\}\big)$ subset of $\mathbb{C}$.

Exercise 3.   Show that if A is any closed bounded subset of $\mathbb{C}$ there is a $T \in \mathcal{L}(\ell_2)$ with $\sigma(T) = A$. [Hint: First show that such an A is *totally bounded* . That is, for $\varepsilon > 0$ there are elements $a_1, \ldots, a_n$, $n = n(\varepsilon)$ in A such that for $a \in A$, $|a - a_j| < \varepsilon$ for some $j = 1, \ldots, n$. Letting $\varepsilon_m = \frac{1}{m}$ use this fact to construct a sequence (a_n) in A with $(\overline{a_n}) = A$, i.e. such an A is separable. Define $T : \ell_2 \to \ell_2$ by $T((\xi_n)) = (a_n \xi_n)$.]

For a set A in $\mathbb{C}$ let $\lambda A = \{\lambda a | a \in A\}$ and $A^n = \{a^n | a \in A\}$.

Exercise 4.   (i) Prove that for $T \in \mathcal{L}(H)$, $\sigma(\lambda T) = \lambda \sigma(T)$.

(ii) Show that $\sigma(T^n) = [\sigma(T)]^n$.

For $T \in \mathcal{L}(H)$ we define

$$r_T = \sup\{|\zeta| : \zeta \in \sigma(T)\}.$$

The number r_T is called the *spectral radius* of T. The above remarks show that $r_T < +\infty$. We now derive the Gelfand-Lorch formula for the spectral radius. This number is important in the subsequent development.

Let $\alpha = e^{\frac{2\pi i}{n}}$. Then $1, \alpha, \alpha^2, \ldots, \alpha^{n-1}$ are the nth roots of unity, that is $\alpha^{nj} = 1$, $j = 0, \ldots, n-1$ (DeMoivre's theorem!) Thus

$$\frac{1}{1 - x^n} = \frac{1}{n}\left(\sum_{i=0}^{n-1}(1 - \alpha^i x)^{-1}\right).$$

(Factor $1 - x^n = \prod_{i=0}^{n-1}(1 - \alpha^i x)$ and use partial fractions). The student should prove a similar statement for operators. We will use this to prove a lemma of Gelfand. The proof is due to Edgar Lorch. Remember C_1 is the unit circle, i.e. $\{z \mid |z| = 1\}$.

V.8 . Let $T \in \mathcal{L}(H)$ and suppose $C_1 \subset \rho(T)$. Let $P = \frac{1}{2\pi i}\int_{C_1} R_\zeta d\zeta$. Then

$$P = \lim_{n\to\infty}(I - T^n)^{-1}.$$

Proof: Let $\alpha^0, \ldots, \alpha^{n-1}$, $\alpha = e^{\frac{2\pi i}{n}}$ be the nth roots of unity. Then

$$\frac{1}{2\pi i}\sum_{j=0}^{n-1}(\alpha^j I - T)^{-1}(\alpha^{j+1} - \alpha^j) = \frac{1}{2\pi i}\sum_{j=0}^{n-1}(I - \alpha^j T)(\alpha - 1)$$

$$= \frac{1}{2\pi i}n(\alpha - 1)(I - T^n)^{-1}.$$

Clearly this is a Riemann-type sum for P and so $P = \lim_{n\to\infty}\frac{1}{2\pi i}n(\alpha - 1)(I - T^n)^{-1}$. But, $n(\alpha - 1) = n\left(e^{\frac{2\pi i}{n}} - 1\right) = n\left[\cos\frac{2\pi}{n} + i\sin\frac{2\pi}{n} - 1\right]$ tends to $2\pi i$ as n tends to $+\infty$ (apply L'Hopital's rule to $\frac{\cos\frac{2\pi}{x} - 1}{\frac{1}{x}}$ and $\frac{\sin\frac{2\pi}{x}}{\frac{1}{x}}$). Thus

$$P = \frac{1}{2\pi i}\int_{C_1} R_\zeta d\zeta = \lim_{n\to\infty}(I - T^n)^{-1}.$$

V.9 Corollary. Suppose that $T \in \mathcal{L}(H)$ has $\rho(T)$ inside C_1. Then $\lim_{n\to\infty}\|T^n\| = 0$.

Proof: Let $r > \max[1, \|T\|]$. By V.6 $\int_{C_r} R_\zeta d\zeta = \int_{C_1} R_\zeta d\zeta$. But, since $r > \|T\|$, $\int_{C_r} R_\zeta d\zeta = 2\pi i I$ and so by V.8 $\lim_{n\to\infty}(I - T^n)^{-1} = I$. Since inversion is continuous (Exercise 1)

$$\lim_{n\to\infty}(I - T^n) = I \quad \text{and so} \quad \lim_{n\to\infty}T^n = 0.$$

We can now prove a very famous result.

V.10 Spectral Radius Theorem. *If* $T \in \mathcal{L}(H)$, $r_T = \lim_{n\to\infty}\|T^n\|^{1/n}$.

Proof: Suppose $S \in \mathcal{L}(H)$ and $r_S < 1$. Then $\sigma(S) \subset C_1$ and so $\|S^n\| \to 0$. Thus $\|S^n\|^{1/n} < 1$ for large n, i.e.,

$$\limsup_{n\to\infty}\|S^n\|^{1/n} \leq 1.$$

Let $T \in \mathcal{L}(H)$ and $\varepsilon > 0$. Also let $S = [r_T + \varepsilon]^{-1}T$. Then $\sigma(S) = \frac{1}{r_T + \varepsilon}\sigma(T)$ (exercise 4) and so $r_S < 1$. Thus $\lim_{n \to \infty} \sup \|S^n\|^{1/n} \leq 1$, so $\lim_{n \to \infty} \sup \|T^n\|^{1/n} \leq r_T + \varepsilon$ and since $\varepsilon > 0$ was arbitrary $\lim_{n \to \infty} \sup \|T^n\|^{1/n} \leq r_T$. Also $r_{T^n} = [r_T]^n$ (again, by exercise 4) and since $r_{T^n} \leq \|T^n\|$, $r_T \leq \|T^n\|^{1/n}$, i.e. $r_T \leq \lim_{n \to \infty} \inf \|T^n\|^{1/n}$.

The spectral radius theorem was first proved (independently) by Gelfand and Lorch.

We remark that the result is true for arbitrary Banach spaces. We haven't used an inner-product anywhere in Chapter V!

Remarks, Exercises, Hints

Spectral theory had it beginnings in the study of matrices on finite dimensional spaces. The use of analytic function theory to extend matrix ideas to operators on Hilbert space, was initiated by several researchers, including Hilbert, Gelfand, Lorch, von-Neumann and F. Riesz. The work of Lorch and (independently) Gelfand appeared in the early forties. Since we do not assume a knowledge of this theory, we have kept the results to a minimum-giving only the results we actually use.

The word "spectrum" is due to Hilbert. Eigenvalue is a strange hybrid, half German, half English (Liverwurst comes to mind!) Despite attempts at translation:"proper value", "characteristic value", even "latent root", this hybrid is evidently here to stay.

The operators we will eventually study have a spectrum which is very easy to describe. However, the spectrum of aribtrary operators can be very complicated. Thus the spectrum has been subdivided in order to study general operators on Hilbert space.

We give a few exercises concerning these subdivisions. These subdivisions are given according to various properties $T \in \mathfrak{L}(H)$. We write for $T \in \mathfrak{L}(H)$ the spectrum of T

$$\sigma(T) = \sigma_c(T) \cup \sigma_r(T) \cup \sigma_p(T)$$

where $\sigma_c(T)$ is the "continuous spectrum": $\lambda \in \sigma_c(T)$ provided $\lambda I - T$ is one-to-one, has a dense range (maybe onto) but the inverse function (defined on the range) is not continuous; $\sigma_r(T)$ is the "residual spectrum": $\lambda \in \sigma_r(T)$ provided $\lambda I - T$ is one-to -one but the range of $\lambda I - T$ is not dense in H; and, $\sigma_p(T)$ in the "point spectrum": $\lambda \in \sigma_p(T)$ provided $\lambda I - T$ is not one-to-one.

1. Consider the operator $T \in \mathfrak{L}(\ell_2)$ discussed in exercise 1 of this chapter (recall that A is an arbitrary closed and bounded subset of $\mathbb{C}$). Show that each a_n (in the sequence constructed) is an eigenvalue and so is in $\sigma_p(T)$. The other points in A are in $\sigma_c(T)$.

2. Consider the following operator on ℓ_2 : $T(x) = \sum\limits_{n=1}^{\infty} (x, e_{n+1}) e_n$, (e_n) the canonical complete orthonormal set in ℓ_2. Show that $\sigma(T) = \{\lambda | |\lambda| \leq 1\}$. Show that if $0 < |\lambda| < 1$ then λ is an eigenvalue of T.

[HINT] Look at the recursive relation implied by $(\lambda I - T)(x) = 0$.

Show that if $|\lambda| = 1$, $\lambda I - T$ is one-to-one, has a dense range, and $(\lambda I - T^{-1})$ is not continuous. Show that if $\lambda = 0$, $\lambda I - T$ is onto but not one-to-one.

Partition the spectrum of T accordingly.

3. Classify the points of $\sigma(T)$ for the operator $T \in \mathcal{L}(\ell_2)$ given by

$$T(x_1, x_2, x_3 \cdots) = (0, x_1, x_2, \cdots).$$

In particular does T have eigenvalues?

We will have much more to say about the operators defined in Exercises 2 and 3 later.

4. Let H be a finite dimensional Hilbert space and $T \in \mathcal{L}(H)$. Show that $\sigma_c(T)$ and $\sigma_r(T)$ are empty.

[HINT] Check your linear Algebra book!)

5. (Scramble operator): Let $T \in \mathcal{L}(\ell_2)$ be defined by

$$T(\alpha_1, \alpha_2, \alpha_3, \alpha_4, \cdots) = (\alpha_2, \alpha_1, \alpha_4, \alpha_3, \cdots).$$

Find $\sigma(T)$. [At some point in your quest consider $S_\lambda \in \mathcal{L}(\ell_2)$ defined for $\lambda \neq 0, \pm 1$ by $S_\lambda (\alpha_1, \alpha_2, \alpha_3 \cdots) = (\alpha_1 + \frac{1}{\lambda}\alpha_2, \alpha_2 + \frac{1}{\lambda}\alpha_1, \cdots).]$

VI. SELF-ADJOINT OPERATORS

An operator $T \in \mathcal{L}(H)$ is *self-adjoint* if $T = T^*$ i.e.,

$$(Tx, y) = (x, Ty)$$

for all $x, y \in H$. If T is self-adjoint (Tx, x) is real for each $x \in H$. Indeed, $(Tx, x) = (x, Tx)$ since T is self-adjoint, but $(Tx, x) = \overline{(x, Tx)}$ by the definition of inner-product. Using this elementary fact we define a partial order on the self-adjoint operators.

VI.1 Definition. If T is a self-adjoint operator we say $T \geq 0$ if and only if $(Tx, x) \geq 0$ for all $x \in H$. For $A, B \in \mathcal{L}(H)$ and self-adjoint we say $A \geq B$ if and only if $A - B \geq 0$.

Exercise 1. (i) $A \geq 0$, $B \geq 0$ implies $A + B \geq 0$;
    (ii) $A \geq 0$ and $\lambda \in \mathbb{C}$, $\lambda \geq 0$ implies $\lambda A \geq 0$;
    (iii) $A \geq B$, $B \geq A$ implies $A = B$; and;
    (iv) $A \geq B$, $B \geq C$ implies $A \geq C$.
    (v) Prove $TT^* \geq 0$ and $T^*T \geq 0$ for any $T \in \mathcal{L}(H)$.

VI.2 Theorem. *If $T \in \mathcal{L}(H)$ is self-adjoint and there is an $\alpha > 0$ such that $\|Tx\| \geq \alpha \|x\|$ for $x \in H$, then T is invertible.*

Proof: By IV.1, $\mathcal{R}(T)$ is closed and from $\|Tx\| \geq \alpha \|x\|$, it follows that T is one-to-one. By IV.2 we need only show that $H = \mathcal{R}(T)$. If there is a $y \in H \backslash \mathcal{R}(T)$ then by exercise 11 chapter IV there is a $z \in H$ such that $(y, z) \neq 0$ and $(Tx, z) = 0$ for all $x \in H$. In particular $(T(Tz), z) = 0$. Since T is self-adjoint $\|Tz\|^2 = (Tz, Tz) = 0$, so $Tz = 0$. Since T is one-to-one, $z = 0$ contradicting $(y, z) \neq 0$.

VI.3 Corollary. If T is self-adjoint and $T \geq I$ the identity operator, T^{-1} exists and $\|T^{-1}\| \geq 1$.

Proof: Since $T \geq I$, $(Tx, x) \geq (x, x) = \|x\|^2$ for all $x \in H$. By the Cauchy-Schwarz-Bunyakovsky inequality it follows that $\|Tx\| \geq \|x\|$. Apply VI.2.

VI.4 Theorem. *If T is self-adjoint then $\sigma(T)$ is real. If also $T \geq 0$, $\sigma(T) \subset [0, +\infty)$.*

Proof: Suppose $\lambda \in \mathbb{C}$ and is not real. Then

$$0 < |\lambda - \bar{\lambda}| \, \|x\|^2 = \left|([T - \lambda I]x, x) - ([T - \bar{\lambda}I]x, x)\right|$$
$$= \left|([T - \lambda I]x, x) - (x, [T - \lambda I]x)\right| \leq 2\|[T - \lambda I](x)\|\|x\|,$$

i.e.,

$$\|[T - \lambda I](x)\| \geq \frac{|\lambda - \bar{\lambda}|}{2}\|x\|$$

for $x \in H$. The same inequality holds if λ and $\bar{\lambda}$ are interchanged. Thus both $T - \lambda I$ and $T - \bar{\lambda}I$ are one-to-one and have closed ranges. Suppose there is a $y \in H\backslash(T - \lambda I)(H)$ with $(y, (T - \lambda I)w) = 0$ for all $w \in H$. Then $((T - \bar{\lambda}I)y, w) = 0$ for all $w \in H$ and so $(T - \bar{\lambda}I)y = 0$. By what was just shown $T - \bar{\lambda}I$ is one-to-one, so $y = 0$. Thus $(T - \lambda I)$ is onto and so $\lambda \in \rho(T)$.

If $T \geq 0$ and $\lambda > 0$ then $T + \lambda I \geq \lambda I$ and so $\lambda^{-1}(T + \lambda I)$ is invertible. Thus $T + \lambda I$ is invertible. That is $-\lambda \in \rho(T)$, i.e.,

$$(-\infty, 0) \subset \rho(T) \quad \text{so} \quad \sigma(T) \subset [0, +\infty).$$

VI.5 Theorem. *If $T \in \mathcal{L}(H)$ is self-adjoint, $\|T^2\| = \|T\|^2$.*

Proof: $\|T^2 x\| = \|T(Tx)\| \leq \|T\|^2\|x\|$ or $\|T^2\| \leq \|T\|^2$ (nothing to do with self-adjointness!). Let $\varepsilon > 0$ and choose $x \in H$, $\|x\| = 1$ with $\|Tx\| \geq \|T\| - \varepsilon$. Then

$$(\|T\| - \varepsilon)^2 \leq \|Tx\|^2 = (Tx, Tx) = (T^2 x, x) \leq \|T^2 x\|\|x\| = \|T^2 x\|$$

by the Cauchy-Schwarz-Bunyakovsky inequality. Thus $\|T\|^2 \leq \|T^2\|$ in this self-adjoint case.

There is a nice and useful corollary to VI.5.

VI.6 Theorem. *If $T \in \mathcal{L}(H)$ is self-adjoint, $r_T = \|T\|$. In particular, in this case, there is a sequence $(\mu_n) \in \sigma(T)$ such that $\lim_{n \to \infty} |\mu_n| = \|T\|$. Moreover $\sigma(T) \subset [-\|T\|, \|T\|]$.*

Proof: By VI.5 $\|T^{2^n}\| = \|T\|^{2^n}$ and so

$$r_T = \lim_{n \to \infty} \|T^n\|^{1/n} = \lim_{n \to \infty} \|T^{2^n}\|^{1/2^n} = \|T\|.$$

The second statement follows from the definition of r_T . Thus for $T \in \mathcal{L}(H)$ and self-adjoint, $\sigma(T) \subset [-\|T\|, \|T\|]$.

Exercise 2. Prove $\sigma(I - T) = \sigma(I) - \sigma(T) = \{1 - \lambda : \lambda \in \sigma(T)\}$.

VI.7 Theorem. *Let $T \in \mathcal{L}(H)$ be self-adjoint. Then $\|T\| \leq 1$ if and only if $-I \leq T \leq I$.*

Proof: Suppose $\|T\| \leq 1$. Then

$$([I - T]x, x) = (x, x) - (Tx, x) \geq \|x\|^2 - \|Tx\|\|x\|$$
$$\geq \|x\|^2 - \|T\|\|x\|^2 = (1 - \|T\|)\|x\|^2 \geq 0,$$

i.e. $I \geq T$. Similarly $T \geq -I$ (check!). Now suppose $-I \leq T \leq I$. Since $\sigma(I-T) \subseteq [0, +\infty)$, $1 - \sigma(T) \subseteq [0, +\infty)$ (exercise) or $\sigma(T) \subseteq (-\infty, 1]$. Also, $\sigma(I + T) \subset [0, +\infty)$ implies

$$\sigma(T) \subseteq [-1, +\infty) \quad \text{so} \quad \sigma(T) \subset [-1, 1].$$

But by VI.6 there is a sequence (μ_n) in $\sigma(T)$ with $(|\mu_n|)$ converging to $\|T\|$. Thus $\|T\| \leq 1$.

We now define the *numerical range* of a self-adjoint operator T.

VI.8 Definition. Let $T \in \mathcal{L}(H)$ be self-adjoint and let $N(T)$ denote the closure of the set $\{(Tx, x) : \|x\| = 1\}$. The set $N(T)$ is called the *numerical range* of T. Clearly, $N(T)$ is closed. (Check!)

Let $m(T) = \inf\limits_{\|x\|=1} (Tx, x)$ and $M(T) = \sup\limits_{\|x\|=1} (Tx, x)$. Clearly $N(T) \subset [m(T), M(T)]$. We are squeezing down on the spectrum of a self-adjoint operator in the following sense.

VI.9 Theorem. *If $T \in \mathcal{L}(H)$ is self-adjoint, $\sigma(T) \subseteq N(T)$.*

Proof: Suppose $\lambda \notin N(T)$. Then $d = \inf\{|\lambda - \mu| : \mu \in N(T)\} > 0$, since $N(T)$ is closed, i.e., $|(Tx, x) - \lambda| \geq d > 0$ for $x \in H$, $\|x\| = 1$. Thus, for $x \in H$, $x \neq 0$, $\left(\frac{x}{\|x\|}, \frac{x}{\|x\|}\right) = 1$, so

$$d \leq \left|\left(T\frac{x}{\|x\|}, \frac{x}{\|x\|}\right) - \lambda\left(\frac{x}{\|x\|}, \frac{x}{\|x\|}\right)\right| = \frac{1}{\|x\|^2} |(Tx, x) - \lambda(x, x)|$$

so $|(Tx - \lambda x, x)| \geq d\|x\|^2$. By the Cauchy-Schwarz-Bunyakovsky inequality $\|T - \lambda I)x\| \geq d\|x\|$. As before, $(T - \lambda I)^{-1}$ exists and $\|(T - \lambda I)^{-1}\| \leq \frac{1}{d}$, i.e. $\lambda \in \rho(T)$ and so $\sigma(T) \subset N(T)$.

There is a nice corollary to VI.6 and VI.9.

VI.10 Theorem.   *If $T \in \mathcal{L}(H)$ is self-adjoint,*

$$\|T\| = \sup_{\|x\|=1} |(Tx, x)| = \max(|m(T)|, |M(T)|).$$

Proof: By the Cauchy-Schwarz-Bunyakovsky inequality

$$\sup_{\|x\|=1} |(Tx, x)| \leq \|T\|.$$

By theorem VI.6 there is a sequence $\mu_n \in \sigma(T)$ such that $\lim_{n\to\infty} |\mu_n| = \|T\|$. By VI.9 $\sigma(T) \subset N(T)$ so each μ_n is of the form $\mu_n = \lim_{m\to\infty} (Tx_m, x_m)$ with $\|x_m\| = 1$.   Thus $\|T\| \leq \sup_{\|x\|=1} |(Tx, x)|$.

We now show that for T self-adjoint $m(T)$ and $M(T)$ are actually in the spectrum of T. The proof we give is due to A. E. Taylor.

Exercise 3.   Let A be a positive and self-adjoint operator. Let $[x, y] = (Ax, y)$. Then $[\bullet, \bullet]$ has all the properties of an inner-product needed to prove the Cauchy-Schwarz-Bunyakovsky inequality, i.e. $[x, y]^2 \leq [x, x][y, y]$. Prove this inequality.

VI.11 Theorem.   *If $T \in \mathcal{L}(H)$ is self-adjoint then $m(T)$ and $M(T)$ are in the spectrum, $\sigma(T)$, of T.*

Proof: We just proved that $M(T)$ and $m(T)$ are finite. Let $\lambda = m(T)$. By the definition of λ, $([T-\lambda I]x, x) \geq 0$ for each $x \in H$. If $[x, y] = ([T-\lambda I]x, y)$ we have by exercise 3 and the Cauchy-Schwarz-Bunyakovsky inequality

$$\begin{aligned}
\|(T - \lambda I)x\|^4 &= \left([T - \lambda I]x, (T - \lambda I)x\right)^2 \\
&= [x, (T - \lambda I)x]^2 \leq [x, x]\left[(T - \lambda I)x, (T - \lambda I)x\right] \\
&= \left((T - \lambda I)x, x\right)\left((T - \lambda I)^2 x, (T - \lambda I)x\right) \\
&\leq \left((T - \lambda I)x, x\right)\|T - \lambda I\|^3 \|x\|^2.
\end{aligned}$$

Thus,

$$\inf_{\|x\|=1} \|(T - \lambda I)x\| \le \inf_{\|x\|=1} ((T - \lambda I)x, x)\|T - \lambda I\|^{3/4} = 0$$

by the definition of λ. However, if $(T - \lambda I)^{-1}$ existed and $\|x\| = 1$ we would have

$$1 = \|x\| \le \|(T - \lambda I)^{-1}\| \|(T - \lambda I)x\|$$

or

$$\|(T - \lambda I)x\| \ge \frac{1}{\|(T - \lambda I)^{-1}\|} > 0.$$

Thus, $\lambda \in \sigma(T)$. In a similar way $M(T) \in \sigma(T)$.

To clarify these ideas consider $T \in \mathfrak{L}(\ell_2)$ defined for $x = (x_i) \in \ell_2$ by $Tx = \sum_{i=1}^{\infty} -\frac{1}{i}x_i e_i$. Then $(Tx, x) = \left(\sum_{i=1}^{\infty} -\frac{1}{i}x_i e_i, \sum_{j=1}^{\infty} x_j e_j \right) = \sum_{i=1}^{\infty} -\frac{1}{i}|x_i|^2$. The student should check that $M(T) = 0$, $m(T) = -1$, $\|T\| = 1$ and, of course, T is self-adjoint.

Let us summarize for self-adjoint T:

(i) $\sigma(T) \subset [m(T), M(T)]$;

(ii) $r_T = \|T\|$;

(iii) $\|T\| = \sup_{\|x\|=1} |(Tx, x)|$; and,

(iv) $m(T), M(T) \in \sigma(T)$.

In particular, $[m(T), M(T)]$ is the smallest interval containing $\sigma(T)$ whenever T is self-adjoint.

Out next goal is to specialize the $T \in \mathfrak{L}(H)$. Then, as we will see, much more can be said.

Remarks, Exercises, and Hints

Hilbert space theory grew out of the work of Hilbert and his school on bilinear and quadratic forms. Although the space ℓ_2 came into being through these studies, Hilbert did not coin the term Hilbert space.

We give a brief discussion of this important topic (bilinear forms) and a few exercises.

Let H be a complex Hilbert space and ϕ a complex valued function on $H \times H$ satisfying $\phi(x, y)$ is linear in x for fixed y and additive and conjugate homogeneous in y for fixed x. Such a ϕ is called a Hermitian bilinear form. If $\psi(x) = \phi(x, x)$, then ψ is the quadratic form associated with ϕ.

1. If ϕ, ψ are as above show that

    a.
    $$\frac{1}{2}[\phi(x, y) + \phi(y, x)] = \psi\left(\frac{x + y}{2}\right) - \psi\left(\frac{x - y}{z}\right)$$

and

    b.
    $$\psi\left(\frac{x + y}{2}\right) - \psi\left(\frac{x - y}{2}\right) + i\psi\left(\frac{x + iy}{2}\right) - i\psi\left(\frac{x - iy}{2}\right) = \phi(x, y).$$

[HINT] Grind it out! The result (b) is called Polarization.

2. Show that if ϕ_1, ϕ_2 are bilinear forms giving rise to the same quadratic form, then

$$\phi_1 = \phi_2.$$

A bilinear form is symmetric provided

$$\phi(x, y) = \overline{\phi(x, y)}.$$

3. A bilinear form is symmetric if and only if the associated quadratic form is real valued.

A bilinear form ϕ is bounded if there is a $K > 0$ such that $|\phi(x, y)| \leq K\|x\| \|y\|$ for all $x, y \in H$. A quadratic form ψ is bounded provided there is an $M > 0$ such that $|\psi(x)| \leq M \|x\|^2$ for all $x \in H$.

We define $\|\phi\| = \inf K$ and $\|\psi\| = \inf M$ where K and M are as indicated.

4. Show that for ϕ and ψ as just described,

$$||\phi|| = \sup\{|\phi(x,y)| : ||x|| = ||y|| = 1\}$$

and

$$\{||\psi|| = \sup|\psi(x)| : ||x|| = 1\}.$$

5. If ψ is the quadratic form associated with the bilinear form ϕ show that ϕ is bounded if and only if ψ is bounded and in the bounded case

$$||\psi|| \leq ||\phi|| \leq 2||\psi||.$$

[HINT] Use 1 (b).

6. Show that, in 5, if ϕ is a bounded, symmetric bilinear form then,

$$||\psi|| = ||\phi||.$$

[HINT] Begin with the real part $\operatorname{Re}\phi(x,y)$. Use 1 to show that

$$\operatorname{Re}\phi(x,y) = \psi\left(\frac{1}{2}(x+y)\right) - \psi\left(\frac{1}{2}(x-y)\right).$$

in the symmetric case. Use the definition of boundedness and the parallelogram identity to show that

$$|\operatorname{Re}\phi(x,y)| \leq \frac{1}{2}||\psi||\left(||x||^2 + ||y||^2\right)|.$$

7. Let $T \in \mathcal{L}(H)$ and let $\phi(x,y) = (Tx,y)$ where $(\bullet,\bullet)$ is the inner product on H. Show that ϕ is a bilinear form and $||\phi|| = ||T||$. (Compare with Exercise 4 in Chapter VI.)

8. If ϕ is a bounded bilinear form show there is a $T \in \mathcal{L}(H)$ with $\phi(x,y) = (Tx,y)$ and $||\phi|| = ||T||$.
[HINT] Think Riesz representation theorem.

9. By using the exercises on bilinear and quadratic forms prove VI.10: If $T \in \mathcal{L}(H)$ is self-adjoint,

$$||T|| = \sup\{|(Tx,x)| : ||x|| = 1\}.$$

The great expositer and polisher, Paul Halmos, stated that no one is interested in Hilbert space but everyone is interested in the operators on Hilbert space.

There is a lot of truth in this. While it is interesting that Hilbert space (and in particular orthogonality in Hilbert space) is unique among Banach spaces it is, in practice, the operators on this wonderful space that demand our attention.

There are two important classes of operators on Hilbert space which will not be used in this work. However, they are of such fundamental importance in the general operator theory on Hilbert space that we devote some exercises to these operators.

An operator $T \in \mathfrak{L}(H)$ is *normal* provided $T^*T = TT^*$ and *unitary* provided

$$T^{-1} \in \mathfrak{L}(H) \text{ and } T^* = T^{-1}.$$

11. Show that if T is self-adjoint or unitary then T is normal.

12. Let I be the identity operator on H. Determine all real $\alpha \neq 0$ such that $T = \alpha i\, I$ is unitary. Do the same with unitary replaced by normal. Give an example of a normal $T \in \mathfrak{L}(H)$ which is not self-adjoint or unitary.

13. Show that if H is a complex Hilbert space, $T \in \mathfrak{L}(H)$ and $(Tx, x) = 0$ for all $x \in H$ then $T = 0$. Show that this assertion is false in the real case.

[HINT] In the complex case compute

$$\big(T(x + y), x + y\big) \text{ and } T\big((ix + y), ix + y\big).$$

In the real case find T so that Tx is orthogonal to x in $\mathbb{R}_2$.

The unitary operators are easy to describe.

14. Show that $T \in \mathfrak{L}(H)$ is unitary if and only if T is an isometric isomorphism.
    [HINT] If T is unitary compute $\|Tx\|^2$. If T is an isometric isomorphism, use 13 to show $T^* = T^{-1}$.

We introduced the shift operator in our brief discussion of the components of the spectrum of an operator. It is worthwhile to look at this important operator and its adjoint again. So let $S \in \mathfrak{L}(\ell_2)$ be given by $S(e_n) = e_{n+1}$ where (e_n) is the canonical orthonormal basis for ℓ_2.

Thus

$$S((\alpha_n)) = \sum_{n=1}^{\infty} \alpha_n \, e_{n+1}.$$

15. (a) Show that $||Sx|| = ||x||$ but S is not onto so S is an isometry but is not unitary.

(b) Show that $\mathcal{R}(S) = \overline{\text{sp}} \{e_n : n \geq 2\}$.

(c) Show that $S^* e_i = e_{i-1}$ for $i > 1$ and $S^* e_1 = 0$. Deduce that

$$\ker S^* = \{x | S^*(x) = 0\} = \text{sp} \{e_1\}.$$

and so, in particular S^{*-1} does not exist.

(d) Show that $\mathcal{R}(S^*) = \ell_2$.

(e) S has a left inverse and S^* a right inverse but SS^* is not one-to-one.

16. Let U denote the set of all unitary operators on Hilbert space. Show that U is a group, i.e. $I \in U$; if $T \in U$ then $T^{-1} \in U$; and, if $S, T \in U$ then $ST \in U$ (and $T^* \in U$ if $T \in U$).

17. If T is unitary, show that $\sigma(T)$ is on the unit circle: $|\lambda| = 1$.

[HINT] Use the fact that T and T^* are isometries to see that λ is in the resolvent of both T and T^* if $|\lambda| > 1$. By definition, $0 \in \rho(T)$. If $0 < |\lambda| < 1, \lambda^{-1} \in \rho(T^*)$. Write $\lambda T (T^* - \lambda^{-1} I) = \lambda I - T$. It should be easy to discover $(\lambda I - T)^{-1}$ now !

As we will see, many of the results concerning self-adjoint operators depend not on the equality $T = T^*$ but on the norm equality $||Tx|| = ||T^* x||$ for each $x \in H$.

18. Show that if $T \in \mathcal{L}(H)$ is self-adjoint and $(Tx, x) = 0$ for all $x \in H$ then $T = 0$, the zero operator (compare with exercise 13 above).

[HINT] From Exercise 4 in Chapter VI deduce that $|(Tx, y)|^2 \leq (Tx, x)(Ty, y)$ or use VI.10.

19. Prove that $T \in \mathcal{L}(H)$ is normal if and only if $||Tx|| = ||T^* x||$ for all $x \in H$.

[HINT] Show that $||Tx|| = ||T^*x||$ if and only if $(x, T^*Tx) = (x, TT^*x)$ for all $x \in H$. If $S = T^*T - TT^*$, S is positive and self-adjoint. Use 18.

20. Show that for any $T \in \mathcal{L}(H)$

$$||T^*T|| = ||T||^2 = ||TT^*||.$$

[HINT] Prove straight forwardly the first equality. Then use T^{**}, T^* for the second.

21. Show that integral powers of normal operators are normal.

22. Show that if $T \in \mathcal{L}(H)$ is normal then $||T||^2 = ||T^2||$. Deduce from this that for normal T, the spectral radius of T is $||T||$. Compare with VI.4 and VI.5.
    [HINT] Use 19 with x replaced with Tx.

There are many interesting relations between (closure of) ranges and kernels of operators and their adjoints on Hilbert space. For instance $\overline{\mathcal{R}(T)}^\perp = \ker T^*$. Indeed $\overline{\mathcal{R}(T)}^\perp = \{y \in H : (Tx, y) = 0 \text{ for all } x \in H\} = \{y \in H : (x, T^*y) = 0 \text{ for all } x \in H\} = \{y \in H : Ty^* = 0\} = \ker T^*$. For $T \in \mathcal{L}(H)$ prove the following:

23. $\overline{\mathcal{R}(T)} = \ker(T^*)^\perp; \overline{\mathcal{R}(T^*)}^\perp = \ker T$; and $, \overline{\mathcal{R}(T^*)} = (\ker T)^\perp$.

24. Show that if T is normal, $\overline{\mathcal{R}(T)}$ and $\ker T$ are orthogonal complements.

25. Show that if T is normal, and $\lambda \in \mathbb{C}$,

$$\overline{\mathcal{R}(\lambda I - T)} = \overline{\mathcal{R}(\bar\lambda I - T^*)}$$

and

$$\overline{\mathcal{R}(\lambda I - T)}^\perp = \ker(\bar\lambda I - T^*).$$

26. Let $T \in \mathcal{L}(H)$ be normal. Show that $\sigma_r(T)$ is empty; show that $\lambda \in \sigma_p(T)$ if and only if $\overline{\mathcal{R}(\lambda I - T)} \neq H$; and identify $\sigma_c(T)$.

27. (Polar decomposition): Show that for $T \in \mathcal{L}(H)$, H complex, the operators $T_1 = \frac{1}{2}(T + T^*)$ and $T_2 = \frac{1}{2i}(T - T^*)$ are self-adjoint. Express T and T^* in terms of T_1 and T_2 and show that the expression is unique.

28. In the decomposition given in 27 show that $T \in \mathcal{L}(H)$ is normal if and only if T_1 and T_2 commute.

29. Show that if $P \in \mathcal{L}(H)$ is a normal projection $(P^2 = P)$, then P is an orthogonal projection i.e. $\mathcal{R}(P) = \ker(P)^{\perp}$.

    [HINT] Use 19.

30. Show that a projection $P \in \mathcal{L}(H)$ is orthogonal if and only if P is self-adjoint. Hence (by 29) a normal projection is self-adjoint.

VII. COMPACT OPERATORS

A set S in a Banach space Y is *compact* provided every sequence (y_n) in S has a subsequence converging to some member of S.

Exercise 1.

(i) (To refresh your memory). If Y is finite dimensional then $S \subset Y$ is compact if and only if S is closed and bounded. (Recall that S is *bounded* in a Banach space Y means there is an $M > 0$ such that $x \in S$ implies $\|x\| \leq M$.)

(ii) show that (i) is false if $Y = \ell_2$.
   [HINT] Let $S = \{x \in \ell_2 \mid \|x\| \leq 1\}$. Then S is closed and bounded. The unit vectors $e_n = (\delta_{in})$ are in S for each n.

A set S in a Banach space Y is *totally bounded* if for $\varepsilon > 0$ there are $x_1, \ldots, x_n$ in S such that $x \in S$ implies there is a j such that $\|x - x_j\| < \varepsilon$. (See Exercise 1, Chapter V.)

Exercise 2. Show that if $S \subset Y$ (Banach space) then S is compact if and only if S is closed and totally bounded.

VII.1 Definition. We say $T \in \mathcal{L}(H)$ is *compact* if T maps the unit ball, $\{x \in H \mid \|x\| \leq 1\}$, into a set whose closure is compact.

This is equivalent to requiring that T map bounded sets into sets with compact closures. (Why?)

Exercise 3. Let (λ_n) be a bounded sequence in $\mathbb{C}$. For $x \in \ell_2$ define
$$Tx = \sum_{n=1}^{\infty} \lambda_n (x, e_n) e_n, \quad (e_n) \text{ the canonical complete orthonormal set for } \ell_2.$$
Then
   (i) $T \in \mathcal{L}(\ell_2)$;
   (ii) T is self-adjoint provided λ_n is real for each n; and,
   (iii) T is compact provided $\lim_{n \to \infty} \lambda_n = 0$.
[HINT (iii) Prove that $S \subset \ell_2$ is compact provided S is closed and for $\varepsilon > 0$ there is an N such that $\left\| \sum_{n=N}^{\infty} (x, e_n) e_n \right\| < \varepsilon$ for all $x \in S$.]

Exercise 4. Show that the closure of the range of a compact operator T is separable.

Let us recall that if $T \in \mathcal{L}(H)$ is self-adjoint then

$$r_T = \|T\| = \max(|m(T)|, |M(T)|) \quad \text{and} \quad m(T), M(T) \in \sigma(T).$$

VII.2 Theorem. *If $T \in \mathcal{L}(H)$ is compact, self-adjoint and non-zero then one of $\|T\|$, $-\|T\|$ is an eigenvalue of T. Moreover, there is an $x \in H$, $\|x\| = 1$, with $|(Tx, x)| = \|T\|$.*

Proof: We know there is a sequence (x_n) in H, $\|x_n\| = 1$ such that

$$\lim_{n \to \infty} (Tx_n, x_n) = a \quad \text{(real) and} \quad |a| = \|T\| \quad \text{(why?).}$$

But,

$$0 \le \|Tx_n - ax_n\|^2 = \|Tx_n\|^2 - 2a(Tx_n, x_n) + a^2\|x\|^2 \le 2a^2 - 2a(Tx_n, x_n).$$

Thus,

$$\lim_{n \to \infty} [Tx_n - ax_n] = 0. \tag{$*$}$$

Since (x_n) is bounded and T compact, there is a subsequence (x_{n_m}) of (x_n) such that (Tx_{n_m}) converges. But then (x_{n_m}) converges by $(*)$ since $\|T\| \ne 0$. Let $\lim_{m \to \infty} x_{n_m} = x$. Then $\|x\| = 1$. By continuity (Tx_{n_m}) converges to Tx and by $(*)$ to ax, i.e. $Tx = ax$ and $|a| = \|T\|$. Since a is real, one of $\pm\|T\|$ is an eigenvalue of T. Also

$$|(Tx, x)| = |(ax, x)| = |a(x, x)| = \|T\|.$$

We are now ready to present one of the main ideas of these lectures.

VII.3 The Procedure. Let $T \in \mathcal{L}(H)$ be compact, self-adjoint and non-zero. Let λ_1 and x_1 be the eigenvalue and (a) corresponding eigenvector given in VII.2. Let $H_0 = H$ and $H_1 = (\text{sp } x_1))^{\perp} = \{x \in H | (x, x_1) = 0\}$. Then H_1 is a closed subspace of H and is hence a Hilbert space. Let T_1 be the restriction of T to H_1. If $x \in H_1$, $(T_1 x, x_1) = (x, Tx_1) = (x, \lambda_1 x_1) = \lambda_1(x, x_1) = 0$ (remember λ_1 is real!). Thus $T_1 x \perp x_1$ so $T_1 x \in H_1$ hence $T_1 \in \mathcal{L}(H_1)$ is compact and self-adjoint. If $T_1 \ne 0$ we can apply VII.2 to T_1 obtaining, λ_2, x_2, $\|x_2\| = 1$, $x_2 \in H_1$, $T_1 x_2 = \lambda_2 x_2$, $|\lambda_2| = \|T_1\|_{H_1} \le \|T\| = |\lambda_1|$ and $(x_1, x_2) = 0$. Let

$$H_2 = [\text{sp}\{x_1, x_2\}]^{\perp} = \{x \in H : (x, x_1) = (x, x_2) = 0\}.$$

Then H_2 is a Hilbert space and if T_2 is T restricted to H_2, $T_2 \in \mathcal{L}(H_2)$ and we can proceed as before. Continuing in this manner we obtain non-zero eigenvalues $\lambda_1, \ldots, \lambda_n$ of T with corresponding norm one eigenvectors $x_1, \ldots, x_n$, subspaces (all Hilbert spaces)

$$H = H_0 \supset H_1 \supset H_2 \supset \cdots \supset H_n$$

with $H_n = [\mathrm{sp}(x_1, \ldots, x_{n-1})]^\perp$ and

$$|\lambda_n| \leq |\lambda_{n-1}| \leq \cdots \leq |\lambda_1|.$$

If this process stops after n steps the range of T is in $\mathrm{sp}\{x_1, \ldots, x_n\}$: for $x \in H$ let $y(x) = x - \sum_{i=1}^{n}(x, x_i)x_i$. Then $(y(x), x_i) = 0$, $i = 1, 2, \ldots, n$ and so $T(y(x)) = 0$; thus,

$$Tx = \sum_{i=1}^{n}(x, x_i)Tx_i = \sum_{i=1}^{n}(x, x_i)\lambda_i x_i.$$

Exercise 5. Let A be an $n \times n$ symmetric matrix. Show that the procedure implies that A can be diagonalized and has a complete orthonormal set of eigenvectors.

We generalize this result and, in a certain sense, obtain the most important result in these lectures. It is a corollary to *The Procedure*.

VII.4 Theorem. (Schur). *Let $T \in \mathcal{L}(H)$ be compact, self-adjoint and non-zero. The Procedure yields a sequence of non-zero eigenvalues $\lambda_1, \lambda_2, \ldots$ and a corresponding orthonormal sequence (x_n) of eigenvectors. If the sequence does not terminate, $\lim_{n \to \infty} |\lambda_n| = 0$. Moreover, for $x \in H$, $Tx = \sum_{n=1}^{\infty} \lambda_n(x, x_n)x_n$. Let G_n be the eigenspace of λ_n, i.e., G_n is the closed linear span of all the x_i's in (x_m) having λ_n as eigenvalue. Then each G_n has finite dimension and its dimension is the number of times λ_n is repeated. Finally the spectrum of T, $\sigma(T)$, is exactly $\{\lambda_n\}$ together with $\{0\}$.*

Proof: Suppose $\lim_{n \to \infty} |\lambda_n| \neq 0$. Then for some $\varepsilon > 0$ there is a subsequence (λ_{n_m}) of (λ_n) with $|\lambda_{n_m}| \geq \varepsilon$. Thus $T\lambda_{n_k}^{-1}x_{n_k} = x_{n_k}$ should have a convergent subsequence since T is compact and $\|\lambda_{n_k}^{-1} x_{n_k}\| \leq \frac{1}{\varepsilon}$. But $\|x_{n_k} - x_{n_\ell}\| = \sqrt{2}$ since (x_n) is an orthonormal set. Thus $\lim_{n \to \infty} |\lambda_n| = 0$.

Clearly (x_n) is a complete orthonormal set in $\overline{\text{sp}\{x_n\}} = \widehat{H}$ and so for $y \in \widehat{H}$, $y = \sum_{n=1}^{\infty} (y, x_n) x_n$ and so by continuity, $Ty = \sum_{n=1}^{\infty} (y, x_n) T x_n = \sum_{n=1}^{\infty} (y, x_n) \lambda_n x_n$. Let $x \in H$. For each n, let $y_n = x - \sum_{i=1}^{n} (x, x_i) x_i$. Then $y_n \in H_n$ and $\|y_n\| \leq \|x\|$ (why?) Since $|\lambda_n| = \|T\|_{H_n}$, the norm of T restricted to H_n,

$$\|Ty_n\| \leq |\lambda_n| \|y_n\| \leq |\lambda_n| \|x\|$$

and so $\lim_{n \to \infty} Ty_n = 0$. It follows (by continuity) that

$$Tx = \sum_{i=1}^{\infty} \lambda_i (x, x_i) x_i.$$

Since the (x_n) are orthonormal (hence linearly independent) the dimension of $\ker(\lambda_n I - T) = \{x \in H | (\lambda_n I - T)x = 0\}$ is greater than or equal to the number of times λ_n is repeated. This number cannot be infinite since if (x_{n_k}) corresponds to λ_n, i.e., $Tx_{n_k} = \lambda_n x_{n_k}$ and $\{n_k\}$ is infinite, we have by compactness of T that for some subsequence $\{n_{k_\ell}\} = \{m\}$ of $\{n_k\}$, $\lim_{m \to \infty} Tx_m = y$ for some $y \in H$. But $Tx_m = \lambda_n x_m$ so (x_m) is convergent, contradicting $\|x_\ell - x_m\| = \sqrt{2}$. Now if the dimension of $\ker(\lambda_n I - T)$ is strictly larger than the number of repeats of λ_n then x_n would be orthogonal to H_i for $i < n$ which is a contradiction (why?). Thus the dimension of $\ker(\lambda_n I - T)$ is the number of repeats of λ_n as an eigenvalue of T. But $\ker(\lambda_n I - T)$ is the eigenspace G_n of λ_n.

Finally, since $\sigma(T)$ is closed and $\lambda_n \to 0$, $0 \in \sigma(T)$, and $\lambda_n \in \sigma(T)$ for each n.

Let $\lambda \in \sigma(T)$. If λ is an eigenvalue of T there is an x_0, $\|x_0\| = 1$ such that $Tx_0 = \lambda x_0$. Then $(\lambda x_0, x_m) = \sum_{n=1}^{\infty} \lambda_n (x_0, x_n)(x_n, x_m) = \lambda_m (x_0, x_m)$, i.e., $\lambda = \lambda_m$ for some m. Since $\lambda I - T$ is not invertible, for each integer m there is a y_m, $\|y_m\| = 1$ such that $\|(\lambda I - T)y_m\| \leq \frac{1}{m}$ (why?) so $\lim_{m \to \infty} (\lambda I - T)(y_m) = 0$. A subsequence (y_{m_n}) has the property that $\lim_{n \to \infty} Ty_{m_n} = y$ exist (since T is compact). But then (λy_{m_n}) must also converge to y. But $0 = \lim_{n \to \infty} (\lambda I - T)(y_{m_n}) = \lambda y - Ty$, i.e., λ is an eigenvalue of T.

For $T \in \mathcal{L}(H)$, $T \neq 0$, compact and self-adjoint, we will call

$$Tx = \sum_{n=1}^{\infty} \lambda_n(x, x_n)x_n, (\lambda_n), (x_n)$$

having the meaning of VII.4 the *Schur representation of T*.

Observe that if $H_0 = \overline{\mathrm{sp}(x_n)}$ then $H = H_0 \oplus H_0^\perp$ and T must be identically 0 on $H_0^\perp$ (why?). Moreover H_0 is clearly separable. Of course H_0 changes with T but because of the Schur representation we can (and do) assume that for compact self-adjoint T, the underlying Hilbert space H is separable.

We have belabored somewhat, the proof of VII.4. However, it is of utmost importance (mentioned earlier) to all the subsequent material.

Thus from now on we *assume* H is separable (thus isometric to ℓ_2) and we change notation and denote the compact operators in $\mathcal{L}(H)$ by $K(H)$. An operator $T \in \mathcal{L}(H)$ is of *finite rank* if $\mathcal{R}(T)$ has finite dimension.

Exercise 6.

(a.) Suppose $T \in \mathcal{L}(H)$ and $\mathcal{R}(T)$ has dimension n. Show that there are x_i, y_i in H with $(x_i, y_j) = \delta_{ij}$ and

$$Tx = \sum_{i=1}^{n} (x, y_i)x_i.$$

(b.) Show that T is of finite rank if and only if T^* is of finite rank.

We need yet another property of compact operators.

VII.5 Theorem. *An operator $T \in K(H)$ if and only if there are finite rank operators A_n such that*

$$\lim_{n \to \infty} \|T - A_n\| = 0.$$

Proof: First suppose such A_n exist. For $\varepsilon > 0$ and (x_n) a bounded sequence in H (say $\|x_n\| \leq M$) there is a N such that $\|T - A_n\| < \frac{\varepsilon}{4M}$ for $n \geq N$. By compactness of A_n some subsequence (x_{n_m}) of (x_n) is such that $(A_n(x_{n_m}))$ converges and is hence Cauchy:

$$\|T(x_{n_m}) - T(x_{n_k})\| \leq \|(T - A_n)(x_{n_m})\| + \|A_n(x_{n_m}) - A_n(x_{m_k})\|$$
$$+ \|(A_n - T)x_{m_k}\| \leq 2\|T - A_n\|M + \|A_n(x_{n_m}) - A_n(x_{n_k})\|$$
$$< \varepsilon$$

if n, n_m, n_k are large. Thus $(T(x_{n_m}))$ is Cauchy and so T is compact.

Now if $T \in K(H)$ and $U = \{x \in H \mid \|x\| \leq 1\}$, $\overline{T(U)}$ is compact. Let (u_i) be a complete orthonormal set for H. Then $x = \sum_{i=1}^{\infty}(x, u_i)u_i$ for $x \in H$ so $Tx = \sum_{i=1}^{\infty}(Tx, u_i)u_i$. Let P_n be the (orthogonal) projection $P_n(x) = \sum_{i=1}^{n}(x, u_i)u_i$ and let $A_n = P_n T$. Then

$$\|T - A_n\| = \sup_{\|x\| \leq 1} \|(T - P_n T)x\| = \sup_{\|x\| \leq 1} \left\| \sum_{i=n+1}^{\infty}(Tx, u_i)u_i \right\|$$

and this expression tends to zero by the next exercise.

Exercise 7. If $K \subset H$ is a compact set and (u_i) a complete orthonormal set for H then for $\varepsilon > 0$ there is a N such that $\left\| \sum_{N}^{\infty}(x, u_i)u_i \right\| < \varepsilon$ for $x \in K$, i.e. the "tails" in the orthogonal expansion of x tend to zero *uniformly* on K. (See Exercise 3(iii) in this chapter.)

Exercise 8.

  (i) Show that $T \in K(H)$ if and only if $T^* \in K(H)$.

  (ii) If $T \in K(H), S \in \mathfrak{L}(H)$, show that TS and $ST \in K(H)$.

  (iii) Show that $T \in K(H)$ if and only if $T^*T \in K(H)$.
       [HINT] If (x_n) is a bounded sequence in H and T^*T is compact, there is a subsequence (z_m) of (x_n) such that (T^*Tz_m) is Cauchy. Compute $\|T[(z_m - z_n)]\|^2]$.

  (iv) Show that $T \in K(H)$ if and only if $TT^* \in K(H)$.

  (v) If $T \in K(H)$ and $T^{-1} \in \mathfrak{L}(H)$, show that H is finite dimensional.

We emphasize (iii): If T is compact T^*T is also compact. Moreover, T^*T is positive and self-adjoint. This fact is very important in what is to follow.

Remarks, Exercises and Hints

The notion of a compact operator is due to Hilbert and F. Riesz. Hilbert's study concerned bilinear forms on ℓ_2. Riesz was the first to systematically study these operators, which he called "vollstetig" and we now translate as "completely continuous." Actually Riesz' definition of what we are calling compact operator, was considerably different (in appearance anyway) from that given in the text. To give Riesz' version of compact we need a definition. A sequence (x_n) in a Hilbert space H is *weakly convergent* to x provided $\lim_{n\to\infty}(x_n,y) = (x,y)$ for each $y \in H$. Riesz' definition of completely continuous operator was (for Hilbert space): $T \in \mathcal{L}(H)$ is completely continuous provided T maps weakly convergent sequences to norm convergent sequences. That is, if (x_n) converges weakly to x in H then $\lim_{n\to\infty} \|Tx_n - Tx\| = 0$.

1. Show that if (x_n) converges weakly to x in H then $\sup \|x_n\| < +\infty$. [HINT] Later you will see that this can easily be killed by the big artillery surrounding the uniform boundedness principle. Since we do not assume knowledge of this principle, we use the route of the "gliding hump principle" which goes back to Lebesgue and Hausdorff. Thus suppose $\lim_{n\to\infty}(x_n,y) = (x,y)$ for each $y \in H$ but $\sup_n \|x_n\| = +\infty$. Choose $f_n \epsilon H, \|f_n\| = 1$ with $|(x_n,f_n)| > \frac{\|x_n\|}{2}$. Let $\alpha(f) = \sup_n |(x_n,f)|$ for each $f \epsilon H$. The weak convergence implies $\alpha(f) < +\infty$ for each f. Determine an increasing sequence of positive integers with

$$\frac{1}{6\cdot 4^k}\|x_{n_k}\| \geq \sum_{i=1}^{k-1} \frac{\alpha(f_{n_i})}{4^i} + k.$$

Let

$$y_0 = \sum_{i=1}^{\infty} \frac{f_{n_i}}{4^i}$$

$$= \left(\sum_{i=1}^{k-1} \frac{f_{n_i}}{4^i} + \frac{f_{n_k}}{4^k} + \sum_{i=k+1}^{\infty} \frac{f_{n_i}}{4^i}\right)$$

$$= y_1 + \frac{f_{n_k}}{4^k} + y_2.$$

Then for each n, $|(x_n,y_1)| \leq \sum_{i=1}^{k-1} \frac{\alpha(f_{n_i})}{4^i}$, $|(x_n,y_2)| \leq \sum_{i=k+1}^{\infty} \frac{\|x_n\|}{4^i} =$

$\frac{1}{3 \cdot 4^k} \| x_n \|$ and so $|(x_{n_k}, y_0)| \geq k$ which contradicts $\lim\limits_{n \to \infty} (x_n, y_0) = (x, y_0)$. This "HINT" is due to G. Köthe.

2. Show that if (x_n) is any bounded sequence in a separable Hilbert space H, a weakly convergent subsequence can be selected from (x_n). [HINT] This will be familiar to some readers as a Cantor diagonalization process. Choose (y_k) dense in H. The sequence $\{(x_{n_1}, y_1)\}$ is bounded in $\mathbb{C}$ (or $\mathbb{R}$) and so has a convergent subsequence, say $\{(x_{n_1}, y_1)\}$. Similarly $\{(x_{n_1}, y_2)\}$ has a convergent subsequence $\{(x_{n_2}, y_2)\}$. Continue this process with the y_is and diagonalize the x_{n_k} to obtain a sequence (z_k) of (x_n) with $\lim\limits_{k \to \infty} (z_k, y_i)$ existing for each i. Use the density of the (y_i) to show that $\{(z_k, x)\}$ is Cauchy for each $x \in H$. Thus

$$f(x) = \lim_{k \to \infty} (x, z_k)$$

defines a continuous linear functional on H (use 1) and so there is a $y \in H$ with $f(x) = (x, y)$.

3. Work with $\overline{sp}\{x_n\}$ and delete the hypothesis that H is separable in 2.

4. Show that $T \in K(H)$ if and only if T is completely continuous in the sense of F. Riesz.
   [HINT] If $T \in K(H)$ and (x_n) is weakly convergent to x then, clearly, (Tx_n) is weakly convergent to Tx. Suppose $\lim\limits_{n \to \infty} \|Tx_n - Tx\| \neq 0$. Then there is a subsequence (Tx_{n_k}) of Tx_n with $\|Tx_{n_k} - Tx\| \geq \epsilon > 0$ for some $\epsilon > 0$. Since (x_n) is bounded (by 1), (Tx_{n_k}) has a convergent subsequence, (Tz_k). If $\lim\limits_{n \to \infty} Tz_n = z$ then $z = Tx$ since (Tz_k) converges weakly to Tx, contradicting $\|Tx_{n_k} - Tx\| \geq \epsilon > 0$. For the converse use 2.

5. (Weighted Shifts) Let (λ_n) be such that $\lambda_n \neq 0$ for each n and $\lim\limits_{n \to \infty} \lambda_n = 0$.

   a. Define $T \epsilon \mathcal{L}(\ell_2)$ by $T(\alpha_1, \alpha_2, \cdots) = (\lambda_1 \alpha_1, \lambda_2 \alpha_2, \cdots \cdots)$. Show that $\sigma_c(T) = \{0\}$. What is $\sigma(T)$?

   b. Shift to the right: $R(\alpha_1, \alpha_2 \cdots) = (0, \lambda_1 \alpha_1, \lambda_2 \alpha_2, \cdots)$. Show that $\sigma_r(R) = \{0\}$. What is $\sigma(T)$?

   c. Shift to the left: $L(\alpha_1, \alpha_2, \cdots) = (\lambda_2 \alpha_2, \lambda_3 \alpha_3, \cdots)$. Show that $\sigma_r(T) = \{0\}$. What is $\sigma(T)$?

6. Show that if $T \in K(H)$, $S \in \mathcal{L}(H)$ and $S^*S \leq T^*T$ then $S \in K(H)$.

7. Show that if $T \in K(H)$ and $\lambda \neq 0$ then $\ker(\lambda I - T)$ has finite dimension. [HINT] If $\ker(\lambda I - T)$ is not finite dimensional, there is an infinite orthonormal set $\{u_n\}$ in $\ker(\lambda I - T)$. The compactness of T implies that $\{u_n\}$ has a convergent subsequence contradicting $||u_n - u_m|| = \sqrt{2}$.

8. Show that if $T \in K(H)$ and $\lambda \neq 0$ then $\mathcal{R}(\lambda I - T)$ is closed. [HINT] Write $H = \ker(\lambda I - T) \oplus \ker(\lambda I - T)^\perp$. Let S be the restriction of $\lambda I - T$ to $\ker(\lambda I - T)^\perp = B$. Then $\ker S = \{0\}$. Since $\mathcal{R}(\lambda I - T) = S(B)$ it suffices to show there is an $\alpha > 0$ such that $||S(b)|| \geq \alpha ||b||$ for $b \in B$. If not there are $b_n \in B$, $||b_n|| = 1$ and $\lim_{n \to \infty} S(b_n) = 0$. Then $\{\lambda b_n\} = \{(\lambda I - T + T)(b_n)\}$ has a convergent subsequence, $\lim_{k \to \infty} \lambda b_{n_k} = b$. Then $b \neq 0$ and $\{S(b_{n_k})\}$ converges to $\lambda^{-1}S(b)$ so $S(b) = 0$. This is a contradiction.

Exercises 7 and 8 play an improtant role in the general theory of compact operators on Hilbert space (and more general Banach spaces).

We will have much more to say about compact operators and certain subclasses of these operators in the remaining chapters.

APPENDIX A: COMPACT INTEGRAL OPERATORS

Recall that $C[a, b]$ denotes the continuous (real or complex) function on $[a, b]$. For $f \in C[a, b]$ define

$$\|f\|_\infty = \sup\{|f(t)| : t \in [a, b]\}.$$

Let $k(s, t)$ be a continuous function on $[a, b] \times [a, b]$ and let K be defined on $C[a, b]$ by

$$K(f)(s) = \int_a^b k(s, t) f(t) dt$$

for $f \in C[a, b]$. The operator K is called a Fredholm integral operator with kernel k. There is a famous theorem due to Ascoli which characterizes compact subsets of $C[a, b]$. Using this result (a statement can be found in any text on functions of a real variable) it is easy to prove that K is a compact operator on $C[a, b]$.

Of course, $C[a, b]$ with $\| \bullet \|_\infty$ is not a Hilbert space. Thus one attempts the obvious: for $f \in C[a, b]$ define

$$\|f\|_2 = \left[\int_b^a |f(t)|^2 dt \right]^{1/2}.$$

Now, $C[a, b]$ with $\| \bullet \|_2$ is an inner product space. Unfortunately, it is not complete. To describe the "completion" requires Lebesgue measure theory and thus gives the student something to look forward to. Actually, on this completion (whatever it is) the operator K will be compact with conditions on the kernel K much weaker than continuity on $[a, b] \times [a, b]$. These integral operators play an important role in certain aspects of the theory of ordinary differential equations and they also have numerous physical applications.

Since we assume no knowledge of Lebesgue theory, we must leave these important operators stranded in this appendix. They will not be explicitly considered or mentioned elsewhere in these notes.

VIII. SQUARE ROOTS

Positive self-adjoint operators, in some sense, play the role in operator theory that the positive numbers play in the real number system. For example, positive real numbers have square roots. Moreover, a positive number has a unique positive square root. Our goal in this chapter is to prove analogous results for positive self-adjoint operators.

If $T \in K(H)$ is non-zero, self adjoint and *positive* we know from the preceding that the spectrum of T, $\sigma(T)$, is contained in the non-negative real axis. Writing the Schur representation $Tx = \sum\limits_{i=1}^{\infty} \lambda_i(x, x_n)x_n$ it follows that $\lambda_i > 0$ for each eigenvalue of T. Let $\mu_i = \lambda_i^{1/2}$ and define $Ax = \sum\limits_{i=1}^{\infty} \mu_i(x, x_i)x_i$. Then A is self-adjoint (check), compact (VII.5) and

$$A^2 x = A \left(\sum_{i=1}^{\infty} \mu_i(x, x_i)x_i \right)$$
$$= \sum_{j=1}^{\infty} u_j \left(\sum_{i=1}^{\infty} u_i(x, x_i)(x_i, x_j)x_j \right)$$
$$= \sum_{i=1}^{\infty} \mu_i^2(x, x_i)x_i = Tx,$$

i.e. $A^2 = T$.

VIII.1 Definition. Suppose $T \in \mathcal{L}(H)$ is self-adjoint and $T \geq 0$. If there is a $A \in \mathcal{L}(H)$, A self-adjoint and $A^2 = T$ then A is called a square root of T. If $A \geq 0$, A is called *the* positive square root of T and we write $A = T^{1/2}$.

We want to show that every positive self-adjoint operator has a *unique* positive square root.

To accomplish this we play on the above analogy to positive real numbers.

First observe that if $S, T \in \mathcal{L}(H)$ are self-adjoint and positive and if $ST = TS$ then TS^2 is positive. We need to show the similar result for TS. Of course if we knew square roots existed we could replace S by $(S^{1/2})^2$ but at this point we know no such thing.

Let us consider another possibility.

Suppose $0 \leq a \leq 1$. Let $a_1 = a$ and $a_{n+1} = a_n - a_n^2$. Then clearly $\sum_{i=1}^{n} a_i^2 = a_1 - a_{n+1} \leq a_1$. Since $\lim_{n \to \infty} a_{n+1} = 0$, $\sum_{i=1}^{\infty} a_i^2 = a_1$. That is, every $a \in [0,1]$ is the *infinite* sum of squares, $\sum_{i=1}^{\infty} a_i^2$, where $a_i \in [0,1]$.

We exploit the operator analog of this fact.

VIII.2 Theorem. *If* $S, T \in \mathcal{L}(H)$ *are self-adjoint and positive and if* $ST = TS$ *then* ST *is positive.*

Proof: If $S = 0$ there is nothing to prove, so suppose $S \neq 0$. Then $\frac{S}{\|S\|} = S_1$ is in the "unit interval" i.e. $0 \leq S_1 \leq I$.

Let $S_{n+1} = S_n - S_n^2$. Observe that $S_{n+1} = S_n^2(I - S_n) + S_n(I - S_n)^2$ so, if $0 \leq S_n \leq I$ for $n > 1$ it follows that $S_{n+1} \geq 0$. But also, $0 \leq I - S_n + S_n^2 = I - S_{n+1}$ so $0 \leq S_n \leq I$ for all $n \geq 1$. As above, $\sum_{i=1}^{n} S_i^2 = S_1 - S_{n+1} \leq S_1$ so $\sum_{i=1}^{n} \|S_i x\|^2 = \sum_{i=1}^{n} (S_i x, S_i x) = \sum_{i=1}^{n} (S_i^2 x, x) \leq (S_1 x, x)$. Thus $\sum_{i=1}^{\infty} \|S_i x\|^2 < +\infty$ for each $x \in H$. In particular $\lim_{n \to \infty} S_n x = 0$ and from $\left(\sum_{i=1}^{n} S_i^2 \right) x = (S_1 - S_{n+1})x$ we obtain $\left(\sum_{i=1}^{\infty} S_i^2 \right) x = S_1 x$. By the continuity of the inner product it follows that TS_1, hence TS is positive.

Exercise 1. If $A, B, T \in \mathcal{L}(H)$ are self-adjoint and $A \leq T \leq B$, show that

$$\|T\| \leq \max\left[\|A\|, \|B\|\right].$$

[HINT] Use VI.7.

Let us return to our analogy. We know from calculus that every bounded monotone sequence of real numbers converges. A similar result holds for self-adjoint operators.

VIII.3 Theorem. *Let* $S, T_n \in \mathcal{L}(H)$ *be self-adjoint and suppose* $T_1 \leq T_2 \leq \cdots \leq T_n \leq S$ *for each* n. *Suppose also that* $ST_n = T_n S$ *for each* n. *Then there is a* $T \in \mathcal{L}(H)$, T *self-adjoint, and* $Tx = \lim_{n \to \infty} T_n x$ *for* $x \in H$.

Proof: Let $S_n = S - T_n \geq 0$. S_n is self-adjoint and for $m < n$,

$$S_m^2 - S_m S_n = (T_n - T_m)(S - T_m) \geq 0$$

by VIII.2, i.e. for $m < n$, $S_m^2 \geq S_{mn}$. Also for $m < n$, $S_m S_n - S_n^2 = (S_m - S_n)S_n = T_n - T_m \geq 0$ so for $x \in H$ and $m < n$,

$$(S_m^2 x, x) \geq (S_m S_n x, x) \geq (S_n^2 x, x) = \|S_n x\|^2.$$

Hence, $\lim_{m \to \infty} (S_m^2 x, x)$ exist [since $((S_m^2 x, x))$ is a monotone decreasing sequence of real numbers which is bounded below!]. Also observe that $-2(S_m S_n x, x) \leq -2(S_n^2 x, x)$ and so

$$\|S_m x - S_n x\|^2 = (S_m^2 x, x) - 2(S_m S_n x, x) + (S_n^2 x, x) \leq (S_m^2 x, x) - (S_n^2 x, x)$$

(check!) thus $(S_n x)$ is Cauchy in norm for each $x \in H$. Since $T_n = S - S_n$, $(T_n x)$ is Cauchy. Let $Tx = \lim_{n \to \infty} T_n(x)$. Then T is linear and

$$\|Tx\| \leq \limsup_{n \to \infty} \|T_n x\| \leq \max \left[\|T_1\|, \|S\|\right] \|x\| \text{ (why?)}$$

i.e., $T \in \mathcal{L}(H)$. Finally, being the point-wise limit of self-adjoint operators, T is self-adjoint.

We now give the main result of section VIII.

VIII.4 Theorem. (Existence of Square Roots). *Every positive self-adjoint $T \in \mathcal{L}(H)$ has a unique positive square root.*

Proof: If $T = 0$ set $A = T^{1/2} = 0$. If $T \neq 0$ let $T_0 = \frac{T}{\|T\|}$. Then $T_0 \leq I$ and if $A^2 = T_0$ then $(\|T\|^{1/2} A)^2 = T$. Thus it suffices to prove the theorem for $T \leq I$. To this end, let $A_0 = 0$ and for $n \geq 0$ let $A_{n+1} = A_n + \frac{1}{2}(T - A_n^2)$.

Observe that $I - A_{n+1} = [I - A_n + \frac{1}{2}(A_n^2 - T)] = \frac{1}{2}[I - 2A_n + A_n^2 + I - T] = \frac{1}{2}[(I - A_n)^2 + (I - T)] \geq 0$ so $A_n \leq I$ for $n \geq 1$.

Clearly $A_1 \geq A_0 = 0$ and if $n > 1$,

$$A_{n+1} - A_n = \left[A_n - \frac{1}{2}(T - A_n^2)\right] - \left[A_{n-1} - \frac{1}{2}(T - A_{n-1}^2)\right]$$

$$= [A_n - A_{n-1}]\left[I - \frac{1}{2}(A_n + A_{n-1})\right] \geq 0$$

so, by induction $A_n \leq A_{n+1}$.

By theorem VIII.2 there is a self-adjoint $A \in \mathcal{L}(H)$ with $\lim\limits_{n \to \infty} A_n x = Ax$ for $x \in H$. Since $A_{n+1} x - A_n x = \frac{1}{2}(Tx - A_n^2 x)$, $\lim\limits_{n \to \infty} [T - A_n^2](x) = 0$ so $A^2 x = Tx$. Clearly A is positive. We claim that A is the unique positive square root of T. Indeed if $B \in \mathcal{L}(H)$ is positive, self-adjoint and $B^2 = A^2 = T$ then from $BT = BB^2 = B^2 B = TB$, it follows that B commutes with A_n for each n and thus $AB = BA$.

For $x \in H$ let $y = (A - B)x$. Then $0 \leq ([A + B]y, y) = ([A^2 - B^2]x, y) = 0$, i.e., (Ay, y) and (By, y) are zero.

By what was just shown, there is a positive, self-adjoint $C \in \mathcal{L}(H)$ with $C^2 = B$. Thus $0 = (By, y) = (C^2 y, y) = \|Cy\|^2$ and so $By = (CCy) = 0$. Likewise $Ay = 0$. Finally $\|[A - B]x\|^2 = ([A - B]y, x) = 0$ and $A = B$.

The proofs of VIII.2, 3, 4 are very similar to those of Riesz and Nagy.

Exercise 2. Find the square roots of the identity on $\mathbb{C}_2$. Which is the unique positive one?

VIII.5 Corollary. Let $T \in \mathcal{L}(H)$. Then T^*T has a unique positive square root $A = (T^*T)^{1/2}$. Moreover, $\|Ax\| = \|Tx\|$ for each $x \in H$ and so in particular,

$$\{x : Ax = 0\} = \{x \mid Tx = 0\}.$$

Proof:

The existence of $A = (T^*T)^{1/2}$ is immediate from VIII.3. Also $\|A^2 x\| = (T^*Tx, x) = \|Tx\|^2$. But A is self-adjoint so $\|A^2 x\| = \|Ax\|^2$.

Exercise 3. Show that if $T \in K(H)$ then $(T^*T)^{1/2} \in K(H)$.
[HINT] See exercise 6 Chapter VII, Notes, Exercises and Hints.

We can now prove one of the principle results in the theory of operators on Hilbert space. For $T \in \mathcal{L}(H)$ remember that $\mathcal{R}(T)$ denotes the range of T. Also we let $A = [T^*T]^{1/2}$.

VIII.6 Polar Decomposition Theorem. *Let $T \in \mathcal{L}(H)$. Then there is a $U \in \mathcal{L}(H)$ such that*

    a) $T = UA$;
    b) $\|Ux\| = \|x\|$ for $x \in \overline{\mathcal{R}(A)}$; and;

c) $Ux = 0$ for $x \in \overline{\mathcal{R}(A)}^{\perp}$.

[An operator U satisfying b) and c) is called a partial isometry].

Proof: Define $V : \mathcal{R}(A) \to \mathcal{R}(T)$ by $V(Ax) = Tx$ for each $x \in H$. Then V is well-defined (by VIII.5) linear and one-to-one (check these statements!). If $y = Ax$, $\|Vy\| = \|VAx\| = \|Tx\| = \|Ax\|$ (by VIII.4) and V is actually an isometry on $\mathcal{R}(A)$. Extend V to $\tilde{V} = \overline{\mathcal{R}(A)} \to \overline{\mathcal{R}(T)}$ by $\tilde{V}y = \lim_{n \to \infty} V(Ax_n)$ where $\lim_{n \to \infty} Ax_n = y$. (Recall that continuous linear operators are uniformly continuous!) Let P be the orthogonal projection of H onto $\overline{\mathcal{R}(A)}$ and let $U = \tilde{V}P$. Pictorially:

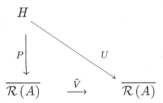

Then for $x \in H$, $UAx = \tilde{V}PAx = \tilde{V}Ax = Tx$ so $UA = T$ and a) is proved. If $x \in \overline{\mathcal{R}(A)}$ there exist $x_n \in H$ with $\lim_{n \to \infty} Ax_n = x$. Then $\|Ux\| = \lim_{n \to \infty} \|UAx_n\| = \lim_{n \to \infty} \|Tx_n\| = \lim_{n \to \infty} \|Ax_n\| = \|x\|$ proving (b).

Since P is 0 on $\overline{\mathcal{R}(A)}^{\perp}$, so is U. This proves c) and we are done.

We need one more small result before getting to our main theorem.

VIII.7 Theorem.   *Let $T \in K(H)$ and let U be the partial isometry of VIII.6. Let (x_n) be the orthonormal eigenvectors of $A = [T^*T]^{1/2}$ (A is compact by exercise 3!) from the procedure. Then (Ux_n) is an orthonormal set.*

Proof: By the properties of U, $\|Ux_n\| = \|x_n\| = 1$. Also, if $m \neq n$, $\|U(x_m - x_n)\|^2 = \|(x_n - x_m)\|^2 = 2$. Thus the real part of (Ux_n, Ux_m) is zero. Similarly, $x_n + ix_m \in \mathcal{R}(A)$ and so, $\|U(x_n + ix_m)\|^2 = \|x_n + ix_m\|^2 = 2$ thus the imaginary part of $(Ux_n, Ux_m) = 0$, i.e. (Ux_n) is an orthonormal set.

Now for the principal result.

VIII.8 The Schmidt Decomposition Theorem. *Let $T \in K(H)$ and let (α_n) denote the eigenvalues of $A = [T^*T]^{1/2}$ given by the procedure. Then there is an orthonormal set (y_n) such that*

$$Tx = \sum_{n=1}^{\infty} \alpha_n(x, x_n)y_n$$

where (x_n) are the orthonormal eigenvectors associated with α_n.

Proof: For $T \in K(H)$, $A = [T^*T]^{1/2} \in K(H)$ is positive and self-adjoint. By the Schur decomposition, $Ax = \sum_{n=1}^{\infty} \alpha_n(x, x_n)x_n$, $\alpha_n \geq 0$ for each n, (x_n) is orthonormal and $Ax_n = \alpha_n x_n$. By VIII.7 if $y_n = Ux_n$ then (y_n) is an orthonormal set. By the Polar decomposition theorem

$$Tx = UAx = U\left(\sum_{n=1}^{\infty} \alpha_n(x, x_n)x_n\right) = \sum_{n=1}^{\infty} \alpha_n(x, x_n)y_n.$$

The (α_n) above are called the *singular* numbers of T. These numbers play an important role in what follows. We will hereafter call the above formula $Tx = \sum_{n=1}^{\infty} \alpha_n(x, x_n)y_n$ with the appropriate meaning for (α_n), (x_n) and (y_n) the *Schmidt* representation of $T \in K(H)$.

In the special case that $T \in K(H)$ is positive and self-adjoint $(T^*T)^{1/2} = (TT)^{1/2}$ so $(TT)^{1/2} = T$ (why?). Since positive square roots are unique the (α_n) above are just the eigenvalues (λ_n) of T. From the polar decomposition theorem $T = UT$, and U is the orthogonal projection to $\overline{R(T)}$. Thus, in this special case, the Schmidt representation coincides with the Schur representation.

Exercise 4. (Important): If $T \in K(H)$ and $\sigma_n(T) = \inf\{\|T - A\| : \text{rank } A < n\}$ (rank A in the dimension of the range of A) show that $\sigma_n(T) = \alpha_n$, where α_n is as above, i.e. the nth (the procedure) eigenvalue of $[T^*T]^{1/2}$.

[HINT] If $T \in K(H)$ write the Schmidt representation $Tx = \sum_{n=1}^{\infty} \alpha_n(x, x_i)y_i$ and let $Ax = \sum_{i=1}^{n-1} \alpha_i(x, x_i)y_i$. Compute $\|Tx - Ax\|^2$ obtaining one inequality. For the more difficult inequality let $A \in \mathcal{L}(H)$ have rank n. Then $Ax = \sum_{k=1}^{n}(x, a_k)b_k$ for some $a_k, b_k, \in H$. Let

$$x_o = \sum_{i=1}^{n} \xi_i x_i$$

where ξ_i is a non-trivial solution of the homogeneous system of equations

$$\sum_{i=1}^{n} \xi_i (x_i, a_k) = 0, k = 1, \cdots n$$

subject to the constraints $\sum_{i=1}^{n} |\xi_i|^2 = 1$. Then $Ax_0 = 0$. Obtain the desired inequality from $||T - A||^2 \geq ||(T - A)x_0||^2$.

Remarks, Exercises, and Hints

The singular numbers of a compact operator T have been studied by numerous researchers. The equality of Exercise 4 was first shown by D. Allalkhverdief in 1957. In Chapter IX we will see the improtant role they play in the study of $T \in K(H)$. The representation of singular numbers given in Exercise 4 makes sense for $T \in \mathcal{L}(H)$ and we define $\sigma_n(T)$ thusly. Here are a few properties the singular numbers enjoy:

1. If $R, S, T \epsilon \mathcal{L}(H)$ show that

    a. $||T|| = \sigma_1(T) \geq \sigma_2(T) \geq \dots$.

    b. $\sigma_{m+n-1}(S + T) \leq \sigma_n(S) + \alpha_m(T)$.

    c. $\sigma_n(RST) \leq ||R|| \, ||T|| \, \sigma_n(S)$.

    d. $\sigma_n(T) = 0$ if $\dim \mathcal{R}(T) < n$.

2. If $I : \ell_2^n \to \ell_2^n$ is the identity operator, show $\sigma_i(I) = 1, i = 1, 2, \cdots, n-1$.

3. Show that $|\sigma_n(S) - \sigma_n(T) \leq ||S - T||$.

4. Show that if T is self adjoint and $A = T^{1/2}$ then $||A|| = ||T||^{1/2}$.

6. Using Exercise 4 in Chapter VI show that if $T \in \mathcal{L}(H)$ is positive and self-adjoint then

$$||Tx|| \leq ||T||^{1/2}(Tx, x)^{1/2}.$$

Thus for a positive self-adjoint operator $T, (Tx, x) = 0$ if and only if $x \in \ker T$.

7. Show that if $S, T \in \mathcal{L}(H)$, S is positive and self-adjoint, and

$$||Sx|| = ||Tx|| \text{ for each } x\epsilon H,$$

then S is the square root of T^*T.

IX. THE WEAK WEYL INEQUALITY

Let us now write $(\lambda_n(T))$ for the eigenvalues of $T \in K(H)$ and $(\sigma_n(T))$ for the singular numbers of T. What, if any, are the relations between $(\lambda_n(T))$ and $(\sigma_n(T))$? This question has been most effectively answered by Herman Weyl. Weyl proved that for each positive integers n and $T \in K(H)$

$$(W) \qquad \prod_{i=1}^{n} |\lambda_i(t)| \le \prod_{i=1}^{n} \sigma_n(T)$$

(where $|\lambda_1| \ge |\lambda_2| \ge \cdots$ as in the Procedure). Using some rather sophisticated integration he proved that (W) implies that for any p, $0 < p < \infty$

$$(\text{weak } W) \qquad \sum_{n=1}^{\infty} |\lambda_n(T)|^p \le \sum_{n=1}^{\infty} \sigma_n(T)^p.$$

Here, we will prove (weak W) using the functional analytic approach of A. Pietsch. This approach utilizes the "factorization of operators techniques" which has proved so successful in modern Banach space theory.

We first introduce some new *Banach* spaces. For $1 \le p < +\infty$ let

$$\ell_p = \{(a_n) \,|\, \sum_{n=1}^{\infty} |a_n|^p < +\infty\}.$$

For $p = +\infty$, $\ell_\infty = \left\{(a_n) \,\middle|\, \sup_n |a_n| < +\infty\right\}$ and c_0 denotes the subset of ℓ_∞ consisting of null sequences:

$$c_0 = \left\{(a_n) \,\middle|\, \lim_{n \to \infty} a_n = 0\right\}.$$

We will show that if one defines $\|(a_n)\|_p = (\sum |a_n|^p)^{1/p}$ for $(a_n) \in \ell_p$, $1 \le p < +\infty$ or $\|(a_n)\| = \sup_n |a_n|$ for $(a_n) \in \ell_\infty$ (or c_0) then ℓ_p, so normed, is a Banach space.

To prove this we need, of course, some new inequalities.

Exercise 1. If $p, q > 1$, $\frac{1}{p} + \frac{1}{q} = 1$ and $f(s,t) = \frac{s^p}{p} + \frac{t^q}{q}$ (s, t real) show that the minimum of $f(s,t)$, subject to the constraint $st = 1$, is 1.

IX.1 Lemma. If $a \ge 0$, $b \ge 0$ then for $p, q > 1$, $\frac{1}{p} + \frac{1}{q} = 1$, $ab \le \frac{a^p}{p} + \frac{b^q}{q}$.

Proof: If one of a, b is 0 the inequality is clearly satisfied. Thus suppose $a > 0$ and $b > 0$. Then

$$\left(\frac{a}{(ab)^{1/p}}\right)\left(\frac{b}{(ab)^{1/q}}\right) = 1$$

and so by exercise 1,

$$\frac{\left(\frac{a}{(ab)^{1/p}}\right)^p}{p} + \frac{\left(\frac{b}{(ab)^{1/q}}\right)^q}{q} \geq 1$$

or

$$ab \leq \frac{a^p}{p} + \frac{b^q}{q}.$$

IX.2 (Hölder's Inequality). Let $p \geq 1$, $q \geq 1$ with $\frac{1}{p} + \frac{1}{q} = 1$ Then for any n, and scalars a_i, $b_i \in \mathbb{C}$

$$\sum_{i=1}^{n} |a_i b_i| \leq \left[\sum_{i=1}^{n} |a_i|^p\right]^{1/p} \left[\sum_{i=1}^{n} |b_i|^q\right]^{1/q}.$$

(One of the sums is replaced by a sup if p or q is ∞.)

Proof: First assume $p > 1$. Let $A = \left[\sum_{i=1}^{n} |a_i|^p\right]^{1/p}$, $B = \left[\sum_{i=1}^{n} |b_i|^q\right]^{1/p}$. If all the a_i or b_i are zero there is nothing to prove; otherwise,

$$\frac{|a_i|}{A}\frac{|b_i|}{B} \leq \frac{\left[\frac{|a_i|}{A}\right]^p}{p} + \frac{\left[\frac{|b_i|}{B}\right]^q}{q} = \frac{|a_i|^p}{pA^p} + \frac{|b_i|^q}{qB^q}$$

by Lemma IX.1. Thus

$$|a_i||b_i| \leq \frac{AB}{pA^p}|a_i|^p + \frac{AB}{qB^q}|b_i|^q.$$

Hence,

$$\sum_{i=1}^{n}|a_i b_i| \leq \frac{AB}{pA^p}\sum_{i=1}^{n}|a_i|^p + \frac{AB}{qB^q}\sum_{i=1}^{n}|b_i|^q$$

$$= \frac{AB}{pA^p}A^p + \frac{AB}{qB^q}B^q$$

$$= AB\left(\frac{1}{p} + \frac{1}{q}\right) = AB$$

and we are done.

If $p = 1$ we adopt the convention that $q = \infty$ and if $p = \infty$, $q = 1$.

Then Hölders inequality becomes (if, e.g. $p = 1$)

$$\sum_{i=1}^{n} |a_i b_i| \leq \left(\sum_{i=1}^{n} |a_i|\right) \sup_i |b_i|.$$

We even need an inequality to see that ℓ_p is a linear space.

IX.3 (Minkowski's inequality).   If $p \geq 1$ then for any n and a_i, $b_i \in \mathbb{C}$

$$\left[\sum_{i=1}^{n} |a_i + b_i|^p\right]^{1/p} \leq \left[\sum_{i=1}^{n} |a_i|^p\right]^{1/p} + \left[\sum_{i=1}^{n} |b_i|^p\right]^{1/p}.$$

[The sums are replaced by sups if $p = \infty$]. In particular, if $a = (a_i) \in \ell_p$ and $b = (b_i) \in \ell_p$ then $a + b \in \ell_p$.

Proof: First suppose $p > 1$. Let q be such that $\frac{1}{p} + \frac{1}{q} = 1$. Then $1 + \frac{p}{q} = p$ and so

$$|a + b|^p = |a + b||a + b|^{p/q} \leq |a||a + b|^{p/q} + |b||a + b|^{p/q}$$

(by the usual triangle inequality). Thus

$$\sum_{i=1}^{n} |a_i + b_i|^p \leq \sum_{i=1}^{n} |a_i||a_i + b_i|^{p/q} + \sum_{i=1}^{n} |b_i||a_i + b_i|^{p/q}$$

$$\leq \left[\sum_{i=1}^{n} |a_i|^p\right]^{1/p} \left[\sum_{i=1}^{n} |a_i + b_i|^p\right]^{1/q} + \left[\sum_{i=1}^{n} |b_i|^p\right]^{1/p} \left[\sum_{i=1}^{n} |a_i + b_i|^p\right]^{1/q}$$

by Hölder's inequality. If all of the $a_i + b_i$ are 0 there is nothing to prove. Otherwise, multiplying the above inequality by $\left(\sum_{i=1}^{n} |a_i + b_i|^p\right)^{-1/q}$ we obtain

$$\left(\sum_{i=1}^{n} |a_i + b_i|^p\right)^{1/p} \leq \left[\sum_{i=1}^{n} |a_i|^p\right]^{1/p} + \left[\sum_{i=1}^{n} |b_i|^p\right]^{1/p}.$$

Here we use $\frac{q}{p} = 1 - \frac{1}{q}$. If $p = 1$, Minkowski's inequality follows from the triangle inequality for scalars. If $p = \infty$ Minkowski's inequality reads

$$\sup_{1 \leq i \leq n} |a_i + b_i| \leq \sup_{1 \leq i \leq n} |a_i| + \sup_{1 \leq i \leq n} |b_i|$$

which again folows from the triangle inequality.

Since Minkowski's inequality is valid for any n, we obtain for $a, b \in \ell_p$, $\|a + b\|_p \leq \|a\|_p + \|b\|_p$. All the other properties of a norm are obvious (check!) for $\| \bullet \|_p$ and so $(\ell_p, \| \bullet \|_p)$ is a normed linear space.

Exercise 2. Prove that ℓ_p, $1 \leq p \leq \infty$ is complete with respect to $\| \bullet \|_p$. [HINT] Mimic the proof that ℓ_2 is complete assuming of course that you proved this fact earlier! Prove that c_0 is *closed* in ℓ_∞ and hence complete.

We now return to compact operators on Hilbert space. We will need an elementary lemma.

IX.4 Lemma. Let $T \in K(H)$ and (e_n) a complete orthonormal set in H. Then $\lim_{n \to \infty} Te_n = 0$.

Proof: If $\lim_{n \to \infty} Te_n \neq 0$ then for some $\varepsilon > 0$ and some subsequence, say (u_n) of (e_n), $\|Tu_n\| > \varepsilon$. Since T is compact there is a subsequence, say (x_n), of (u_n) such that $\lim_{n \to \infty} Tx_n = x$ exists. Thus

$$\lim_{n \to \infty} (x_n, T^*x) = \lim_{n \to \infty} (Tx_n, x) = (x, x).$$

But since (e_n) is a complete orthonormal set

$$T^*x = \sum_{k=1}^{\infty} (T^*x, e_k) \, e_k.$$

In particular, $\lim_{k \to \infty} (T^*x, e_k) = 0$. But then $\lim_{n \to \infty} (T^*x, x_n) = 0$ since (x_n) is a subsequence of (e_k). It follows that $(x, x) = 0$ so $x = 0$. But (x_n) is a subsequence of (u_n) and this contradicts $\|Tu_n\| \geq \varepsilon$. Thus $\lim_{n \to \infty} Te_n = 0$.

We now reduce the question of compactness to a problem concerning the canonical complete orthonormal set (e_n) of ℓ_2. This characterization is due to A. Pietsch.

IX.5 Theorem. *Let $T \in \mathcal{L}(H)$. Then $T \in K(H)$ if and only if $((TUe_n, Ve_n)) \in c_0$ for every $U, V \in \mathcal{L}(\ell_2, H)$, the bounded linear operators from ℓ_2 to H.*

Proof: Suppose $((TUe_n, Ve_n)) \in c_0$ for all $U, V \in \mathcal{L}(\ell_2, H)$. Let $\varepsilon > 0$ and choose orthonormal sets (x_n), (y_n) maximal with respect to the property:

$(Tx_n, y_n) \geq \varepsilon$. (How is this possible?) Let (x_n) and (y_n) be indexed by the set Γ. We claim that Γ is finite. If not, let $U(\xi_n) = \sum\limits_{n=1}^{\infty} \xi_n x_n$ and $V(\eta_n) = \sum\limits_{n=1}^{\infty} \eta_n y_n$. Then $U, V \in \mathfrak{L}(\ell_2, H)$ (why?) and $\lim\limits_{n \to \infty} (Tx_n, y_n) = \lim\limits_{n \to \infty} (TUe_n, Ve_n) = 0$ by hypothesis. Thus only finitely many terms can be bounded away from zero, so Γ is finite. Let $Ax = \sum\limits_{i \in \Gamma}(x, x_i)x_i$, $Bx = \sum\limits_{i \in \Gamma}(x, y_i)y_i$. We claim that $\|(I - B)T(I - A)\| \leq \varepsilon$ (I the identity on H). Indeed, if not, there exists $x, y \in H$ with

$$\left| ((I - B)T(I - A)x, y) \right| > \varepsilon \|x\| \|y\|. \quad \text{(why?)}$$

It is easy to check that $I - A$ and $I - B$ are self-adjoint. Let $x_0 = \frac{x - Ax}{\|x - Ax\|}$, $y_0 = \frac{y - By}{\|y - By\|}$ (clearly, neither $x - Ax = 0$ or $y - By = 0$). Then $(Tx_0, y_0) > \varepsilon$ and

$$(x_i, x_0) = \frac{1}{\|x - Ax\|}(x_i, (I - A)x) = \frac{1}{\|x - Ax\|}((I - A)x_i, x) = 0.$$

Similarly $(y_i, y_0) = 0$. This contradicts the maximality of (x_i), (y_i). Thus $\|(I - B)T(I - A)\| < \varepsilon$. Thus T is the limit of finite rank operators and so $T \in K(H)$.

Now suppose $T \in K(H)$, $U, V \in \mathfrak{L}(\ell_2, H)$. Define $V^* : H \to \ell_2$ by (V^*h, x) (inner product in ℓ_2) $= (h, Vx)$ (inner product in H). Then V^* is continuous (why?) and so V^*TU is compact. Thus by lemma IX.4, $\lim\limits_{n \to \infty} V^*TUe_n = 0$. But $|(TUe_n, Ve_n)| = |(V^*TUe_n, e_n)| \leq \|V^*TUe_n\|$ by the Cauchy-Schwarz-Bunyakovsky inequality. This proves the theorem. Observe that V^* is the Banach space adjoint operator in this case.

Recall that the singular numbers $(\sigma_n(T))$ of $T \in K(H)$ are the eigenvalues of $(T^*T)^{1/2}$ and by a previous exercise are given by $\sigma_n(T) = \inf \|T - A\|$ where the infimum is taken over all operators $A \in \mathfrak{L}(H)$ with rank A (the dimension of the range of A) $< n$.

Exercise 3.   Show that $T \in K(H)$ if and only if $\lim\limits_{n \to \infty} \sigma_n(T) = 0$ i.e. $(\sigma_n(T)) \in c_0$. Let

$$S_p(H) = \{T \in K(H) | (\sigma_n(T)) \in \ell_p\}, \quad 1 \leq p < \infty.$$

These are the Schatten p-classes, named in honor of Robert Schatten who essentially formulated these definitions.

We give a characterization of $\mathcal{S}_p$-operators, also due to Pietsch, completely analogous to that of compact operators given in IX.5.

Exercise 4. Let $T \in \mathcal{S}_p(H)$, $A \in \mathfrak{L}(\ell_2, H)$, $B \in \mathfrak{L}(H, \ell_2)$. Then $ATB \in \mathcal{S}_p(\ell_2)$.

IX.6 Theorem. *Let $T \in \mathfrak{L}(H)$. Then $T \in \mathcal{S}_p(H)$ if and only if for every $U, V \in \mathfrak{L}(\ell_2, H)$, $((TUe_n, Ve_n)) \in \ell_p$.*

Proof: Suppose the condition holds. Then by IX.5 $T \in K(H)$ and thus has a Schmidt representation

$$Tx = \sum_{n=1}^{\infty} \sigma_n(T)(x, x_n)y_n, \quad (x_n), (y_n)$$

orthonormal sets. Let $U((\xi_i)) = \sum_{i=1}^{\infty} \xi_i x_i$ and $V((\xi_i)) = \sum_{i=1}^{\infty} \xi_i y_i$. Then U and V are continuous and $(TUe_i, Ve_i) \in \ell_p$. But $(TUe_i, Ve_i) = (Tx_i, y_i) = (\sigma_i(T)y_i, y_i) = \sigma_i(T)$. Thus $T \in \mathcal{S}_p(H)$.

The proof in the other direction is considerably more difficult and requires judicious use of Hölder's inequality. Bear with me! Suppose $T \in \mathcal{S}_p(H)$, $U, V \in \mathfrak{L}(\ell_2, H)$. Let $\frac{1}{p} + \frac{1}{q} = 1$ and let V^* be defined as in IX.5. Let $T_0 = V^*TU$. By exercise 4, $T_0 = V^*TU \in \mathcal{S}_p(\ell_2)$. Thus T_0 has a Schmidt representation

$$T_0 x = \sum_{n=1}^{\infty} \sigma_n(T_0)(x, x_n)y_n$$

for some orthonormal sets (x_n) and (y_n). Thus,

$$\sum_{i=1}^{\infty} \sigma_i(T_0) |(x_i, e_n)|^2 = \sum_{i=1}^{\infty} \sigma_i(T_0) |(x_i, e_n)|^{2/p}|(x_i, e_n)|^{2/q}$$

$$\leq \left[\sum_{i=1}^{\infty} \sigma_i(T_0)^p|(x_i, e_n)|^2\right]^{1/p} \left[\sum_{i=1}^{\infty} |(x_i, e_n)|^2\right]^{1/q}$$

by Hölder's inequality. Thus

$$\left[\sum_{i=1}^{\infty} \sigma_i(T_0)|(x_i, e_n)|^2\right]^{p/2} \leq \left[\sum_{i=1}^{\infty} \sigma_i(T_0)^p|(x_i, e_n)|^2\right]^{1/2} \left[\sum_{i=1}^{\infty} |(x_i, e_n)|^2\right]^{p/2q}$$

$$\leq \left[\sum_{i=1}^{\infty} \sigma_i(T_0)^p|(x_i, e_n)|^2\right]^{1/2}$$

since $\sum_{i=1}^{\infty} |(x_i, e_n)|^2 \le \|e_n\|^2 = 1$ (by Bessel's inequality). Similarly,

$$\left[\sum_{i=1}^{\infty} \sigma_i(T_0) |(y_i, e_n)^2 \right]^{p/2} \le \left[\sum_{i=1}^{\infty} \sigma_i(T_0)^p |(y_i, e_n)|^2 \right]^{1/2}.$$

Therefore,

$$|(T_0 e_n, e_n)| = \left| \sum_{i=1}^{\infty} \sigma_i(T_0)^{1/2} \sigma_i(T_0)^{1/2} (e_n, x_i)(y_i, e_n) \right|$$

$$\le \left[\sum_{i=1}^{\infty} \sigma_i(T_0) |(x_i, e_n)|^2 \right]^{1/2} \left[\sum_{i=1}^{\infty} \sigma_n(T_0) |(y_i, e_n)|^2 \right]^{1/2},$$

again by Hölder's inequality.

Hence,

$$|(T_0 e_n, e_n)|^p \le \left[\sum_{i=1}^{\infty} \sigma_i(T_0) |(x_i, e_n)|^2 \right]^{p/2} \left[\sum_{i=1}^{\infty} \sigma_i(T_0) |(y_i, e_n)|^2 \right]^{p/2}$$

$$\le \left[\sum_{i=1}^{\infty} \sigma_i(T_0)^p |(x_i, e_n)|^2 \right]^{1/2} \left[\sum_{i=1}^{\infty} \sigma_i(T_0)^p |(y_i, e_n)|^2 \right]^{1/2}.$$

Finally,

$$\sum_{n=1}^{\infty} |(T_0 e_n, e_n)|^p \le \sum_{n=1}^{\infty} \left[\sum_{i=1}^{\infty} \sigma_i(T_0)^p |(x_i, e_n)|^2 \right]^{1/2}$$

$$\left[\sum_{i=1}^{\infty} \sigma_i(T_0)^p |(y_i, e_n)|^2 \right]^{1/2}$$

$$\le \left\| \sum_{i=1}^{\infty} \sigma_i(T_0)^{p/2} x_i \right\| \left\| \sum_{i=1}^{\infty} \sigma_i(T_0)^{p/2} y_i \right\|$$

$$\le \left(\sum_{i=1}^{\infty} \sigma_i(T_0)^p \right)^{1/2} \left(\sum_{i=1}^{\infty} \sigma_i(T_0)^p \right)^{1/2} = \sum_{i=1}^{\infty} \sigma_i(T_0)^p.$$

All the above is because (e_n) is a complete orthonormal set in ℓ_2, and (x_i), (y_i) are orthonormal in H. Thus $(T_0 e_n, e_n) = (TU e_n, V e_n) \in \ell_p$.

Exercise 5.   For $1 \le p < \infty$ and $T \in \mathcal{S}_p(H)$ let $s_p(T) = \left(\sum_{n=1}^{\infty} \sigma_n(T)^p \right)^{1/p}$. Then $(\mathcal{S}_p, s_p)$ is a Banach space. Moreover $s_p(T) =$

$\sup \{\sum |(TUe_n, Ve_n)|^p\}^{1/p}$ where the sup is taken over all $U, V \in \mathfrak{L}(\ell_2, H)$ with $\|U\| = \|V\| = 1$.

We need an algebraic result before proving the "weak form" of the Weyl inequality.

IX.7 Theorem. *Let $T \in K(H)$ with eigenvalues (λ_n) ordered so that $|\lambda_1| \geq |\lambda_2| \geq \cdots$ (counting multiplicities). Then for each n there is an orthonormal set $x_1, \ldots, x_n$ such that $(Tx_i, x_i) = \lambda_i$ for $i = 1, \ldots, n$.*

Proof: (Dunford and Schwartz). Let z_i be an eigenvector corresponding to λ_i for each $i = 1, \ldots, n$, let $E_n = \mathrm{sp}\{z_i \mid 1 \leq i \leq n\}$ and let T_n be the restriction of T to E_n. Then T_n has eigenvalues $\lambda_1, \ldots, \lambda_n$ and $T_n(E_n) \subset E_n$. Hence T_n has a representation as an $n \times n$ matrix. We claim there is an orthonormal basis (x_i) for E_n for which the matrix of T_n is subdiagonal. This claim is proved by induction. If $n = 1$ there is nothing to prove. Let $n > 1$. The induction hypothesis is that if X has dimension $n - 1$ and $T_n X \subset X$ there is an orthonormal set $x_1, \ldots, x_{n-1}$ with T_n represented by (a_{ij}) where $a_{ij} = (Tx_i, x_j)$ and $a_{ij} = 0$ if $j > i$ (this is the meaning of subdiagonal!)

Let λ be an eigenvalue of T_n and $S_0 = (T_n - \lambda I)E_n$. Now $S_0 \neq E_n$ since $\{x \mid (T_n - \lambda I)(x) = 0\} \neq \{0\}$. Thus the dimension of S_0 is no greater than $n - 1$. Let $S \supseteq S_0$ be an $(n - 1)$ dimensional subspace of E_n. Then if $s \in S$, $(T_n - \lambda I)s = s_0 \in S_0$ and so $T_n s = \lambda s + s_0 \in S$, i.e. $T_n(S) \subset S$. By the induction hypothesis there are orthogonal vectors $x_1, \ldots, x_{n-1}$ with the desired properties. In particular $([T - \lambda I]x_i, x_j) = 0$ for $j > i$. Extend $x_1, \ldots, x_{n-1}$ to an orthonormal basis for E_n. Then the determinant

$$\det(\lambda I - T_n) = \prod_{i=1}^{n} [\lambda - (T_n x_i, x_i)].$$

The eigenvalues of T_n are the roots of $\det(\lambda I - T_n)$ and so $\lambda_i = (T_n x_i, x_i)$.

We can now prove the weak Weyl inequality.

IX.8 (Weak Weyl inequality). Let $1 \leq p < \infty$, and $T \in K(H)$. Then

$$\left(\sum_{n=1}^{\infty} |\lambda_n(T)|^p \right)^{1/p} \leq \left(\sum_{n=1}^{\infty} \sigma_n(T)^p \right)^{1/p}.$$

In particular, if $T \in \mathcal{S}_p(H)$,

$$\left(\sum_{n=1}^{\infty} |\lambda_i(T)|^p \right)^{1/p} \le s_p(T).$$

Proof: Fix n. Let $x_1, \dots, x_n$ be determined by IX.7 so that $(Tx_i, x_i) = \lambda_i$ and (x_i) is orthonormal. Define $U, V \in \mathcal{L}(\ell_2, H)$ by

$$U(e_i) = V(e_i) = x_i \quad i \le n$$
$$U(e_i) = V(e_i) = 0 \quad i > n.$$

Then $\lambda_i = (Tx_i, x_i) = (TUe_i, Ve_i)$. Thus by IX.6 and exercise 5

$$\left(\sum_{i=1}^{n} |\lambda_i|^p \right)^{1/p} = \left(\sum |(TUe_i, Ve_i)|^p \right)^{1/p}$$

$$\le \sup_{\|U\|=\|V\|=1} \left(\sum_{i=1}^{\infty} (TUe_i, Ve_i)^p \right)^{1/p}$$

$$= \left(\sum \sigma_i(T)^p \right)^{1/p} = s_p(T).$$

APPENDIX B: THE WEYL INEQUALITY

We have given the proof of the *weak* form of the Weyl inequality in order to exploit some Banach space techniques. Moreover, this form is all we need for what follows. Also to obtain this weak form from the multiplicative form requires, as mentioned, some tricky integration techniques.

However, the multiplicative form is not too difficult to prove and could be used in chapter X for the "localization of Eigenvalues" formula.

Of course, by this time the reader should suspect that yet another inequality will be used to prove the Weyl inequality.

B.1 Hadamard's Inequality.   If (a_{ij}) is an $n \times n$ matrix of complex numbers then

$$(\text{H}) \qquad |\det(a_{ij})| \leq \prod_{j=1}^{n} \left\{ \sum_{i=1}^{n} |a_{ij}|^2 \right\}^{1/2}.$$

Proof: (Dunford and Schwartz).   The proof is by induction. There is nothing to prove if $n = 1$. Suppose the result to be true for $n - 1$ and let (a_{ij}) be an $n \times n$ matrix. Write

$$u_j = \begin{pmatrix} a_{1j} \\ \vdots \\ a_{nj} \end{pmatrix}, \quad j = 1, \ldots, n.$$

If $u_j = 0$ (as an element of $\mathbb{C}_n$) there is nothing to prove. Since multiplying both sides of (H) by the same non-zero constant does not affect the validity of (H) we may assume that

$$\|u_1\| = \left(\sum_{j=1}^{n} |a_{j1}|^2 \right)^{1/2} = 1$$

(this is the norm of u_1 in ℓ_2^n).

Let $\{v_1, \ldots, v_n\}$ be an orthonormal basis for ℓ_2^n with $v_1 = u_1$ and define $W : \ell_2^n \to \ell_2^n$ by $W(v_k) = e_k$, where $e_k = (\delta_{ik})_{i=1}^n$. Clearly $|\det W| = 1$

(why?) Let $W u_k = (w_{1k}, \ldots, w_{nk})$. Then

$$|\det(a_{ij})| = |\det(u_1, u_2, \ldots, u_n)| = |\det W||\det(u_1, \ldots, un)|$$

$$= |\det(W u_1, \ldots, W u_n)| = \det \begin{vmatrix} 1 & w_{12} & \cdots & w_{1n} \\ 0 & w_{22} & \cdots & w_{2n} \\ \vdots & \vdots & \vdots & \vdots \\ 0 & w_{n2} & \cdots & w_{nn} \end{vmatrix}$$

$$= \det \begin{vmatrix} w_{22} & \cdots & w_{2n} \\ \vdots & & \vdots \\ w_{n2} & \cdots & w_{nn} \end{vmatrix}.$$

Thus by the induction hypothesis,

$$|\det(a_{ij})| \leq \prod_{j=2}^{n} \left(\sum_{i=2}^{n} |w_{ij}|^2 \right)^{1/2}.$$

But

$$\left(\sum_{i=2}^{n} |w_{ij}|^2 \right)^{1/2} \leq \left(\sum_{i=1}^{n} |w_{ij}|^2 \right)^{1/2} = \|W u_j\| = \|u_j\|, \quad j = 2, \ldots, n$$

(why?) Thus, since $\|u_1\| = 1$, (H) holds.

B.2 Corollary. Let $A \in \mathcal{L}(\ell_2^n)$ have matrix representation (a_{ij}). Then

$$|\det(a_{ij})| \leq \prod_{k=1}^{n} \sigma_k(A),$$

where $\sigma_k(A)$ are the singular numbers of A.

Proof: Let $Ax = \sum_{k=1}^{n} \sigma_k(A)(x, x_k) y_k$ be the Schmidt representation of A where (x_i), (y_i) are orthonormal in ℓ_2^n. Let $(e_k)_{i=1}^{n}$ be the canonical complete orthonormal set of ℓ_2^n as before (k^{th} unit vector) then

$$a_{ij} = \sum_{k=1}^{n} (e_i, x_k) \sigma_k(A)(y_k, e_j)$$

so

$$|\det(a_{ij})| = |\det(e_i, x_k) \prod_{k=1}^{n} \sigma_k(A) \det(y_k, e_j)|. \qquad \text{(why?)}$$

By Hadamard's inequality

$$|\det(e_i, x_k)| \le \prod_{k=1}^{n} \left(\sum_{i=1}^{n} |(e_i, x_k)|^2 \right)^{1/2} = \prod_{k=1}^{n} \|x_k\| = 1.$$

Similary $|\det(e_i, x_k)| \le 1$ so $|\det(a_{ij})| \le \prod_{k=1}^{n} \sigma_k(A)$.

B.3 Weyl's Inequality.   Let $T \in K(H)$ have eigenvalues $(\lambda_n(T))$ arranged $|\lambda_1(T)| \ge |\lambda_2(T)| \ge \cdots$. Then for any n

$$\prod_{k=1}^{n} |\lambda_k(T)| \le \prod_{k=1}^{n} \sigma_k(T).$$

Proof: Fix n and let $\lambda_1(T), \ldots, \lambda_n(T) \ne 0$ have eigenvectors $x_1, \ldots, x_n$. Let $E_n = \mathrm{sp}\{x_1, \ldots, x_n\}$. Let $J : E_n \to H$ be the inclusion map. Let $\{y_1, \ldots, y_n\}$ be an orthonormal basis for E_n and extend (y_i) to a complete orthonormal set (z_i) for H. Let Q be the orthogonal projection onto E_n. Finally let $\phi : E_n \to \ell_2^n$ be the isometry $\phi(y_i) = e_i$, $i = 1, \ldots, n$. Then $\phi Q T J \phi^{-1} \in \mathcal{L}(\ell_2^n)$ and has eigenvalues $\lambda_1(T), \ldots, \lambda_n(T)$ (check!). Thus

$$\prod_{k=1}^{n} |\lambda_k(T)| = \prod_{k=1}^{n} |\lambda_k(\phi Q T J \phi^{-1})| = |\det \phi Q T J \phi^{-1}| \le \prod_{k=1}^{n} \sigma_k(\phi Q T J \phi^{-1})$$

$$\le \prod_{k=1}^{n} \|\phi\| \, \|\phi^{-1}\| \, \|Q\| \, \|J\| \sigma_k(T) = \prod_{k=1}^{n} \sigma_k(T).$$

Picture: $\ell_2^n \xrightarrow{\phi^{-1}} E_n \xrightarrow{J} H \xrightarrow{T} H \xrightarrow{Q} E_n \xrightarrow{\phi} \ell_2^n$

Remarks, Exercises and Hints

As seen in Appendix B, the Weyl inequality follows rather easily from the Hadamard inequality. Weyl proved this result in the early forties using completely different ideas. But, as mentioned at the beginning of Chapter IX, what we call the *weak Weyl inequality* does not just fall out of the Weyl inequality. It requires work!. Moreover, as we will see, it certainly is not a *weak* result. The Banach space approach was deliberately chosen in order to introduce some Banach space techniques. It also allows for some exercises to give an indication of what happens in operator theory if the underlying space is not a Hilbert space.

1. Show that the conjugate space, $(\ell_1)^*$ of ℓ_1 can be identified with ℓ_∞.
   [HINT] Let u_k be the sequence in ℓ_1 whose entries are 1 in the $k^{\underline{th}}$ entry and 0 elsewhere. Clearly, if $\lambda = (\lambda_k) \in \ell_1$ then $\lambda = \sum\limits_{k=1}^{\infty} \lambda_k u_k$. If $f \in \ell_1^*$, let $\alpha_k = f(u_k)$.

Let

$$\left\{ \begin{array}{c} \beta_k = \mathrm{sgn}\ \bar{\alpha}_k\ \text{if}\ k = n \\ 0\ \text{if}\ k \neq n. \end{array} \right\}$$

Then

$$\beta = (\beta_k) \in \ell_1\ \text{and}\ ||\beta|| = 1.$$

Thus,

$$f(\beta) = \Sigma \beta_k \alpha_k = |\alpha_k|$$

so

$$|\alpha_n| \leq ||f||\,||\beta|| \leq ||f||\ \text{and}\ (\alpha_n) \in \ell_\infty$$

and

$$||\alpha||_\infty \leq ||f||.$$

On the other hand, if $\alpha = (\alpha_n) \in \ell\infty$ and $\lambda = (\lambda_n) \in \ell_1$ then $f(\lambda) = \Sigma \alpha_n \lambda_n$ clearly defines an $f \in \ell^*$, and $||f|| \leq ||\alpha||_\infty$. Part of this exercise is to explain all the clearlys!

We say that a sequence (x_n) in a Banach space X converges weakly to $x \in X$ provided $\lim\limits_{n \to \infty} f(x_n) = f(x)$ for each $f \in X^*$. (See the introductory remarks in the notes to VII.)

2. Show that in ℓ_1 a sequence (x_n) converges weakly to x if and only if

$$\lim_{n\to\infty} \|x_n - x\| = 0.$$

[HINT] The non-trivial implication can be proved using another gliding hump argument. If (x_n) converges weakly to x in ℓ_1 by subtracting x we can assume (x_n) converges weakly to 0. If (x_n) does not converge to 0 in the ℓ_1 norm we can construct a sequence $\alpha = (\alpha_n) \in \ell_\infty$ such that

$$\lim_{n\to\infty} \alpha(x_n) \neq 0, \text{ where } \alpha(\beta) = \sum_{i=1}^{\infty} \alpha_i \beta_i, \text{ for } (\beta_i = \beta \in \ell_1.$$

Using Exercise 1 yields a contradiction. The following *hint* is from the master himself - S. Banach: Let $x_n = (x_i^n)$ and suppose $\lim\limits_{n\to\infty} x_n \neq 0$ in ℓ_1. Then there is $\epsilon > 0$ and a sequence n_j of positive integers such that

$$\|x_{n_j}\| = \sum_{i=1}^{\infty} |x_i^{n_j}| > \epsilon.$$

Choose N_1 large enough so that

$$\sum_{N_1+1}^{\infty} |x_i^{n_1}| < \frac{\epsilon}{5}$$

and

$$\sum_{i=1}^{N_1} |x_i^{n_1}| > \frac{4\epsilon}{5}.$$

Choose numbers $\alpha_1 \cdots \alpha_{N_1}$ with $|\alpha_i| = 1$ $i = 1, \cdots N_1$ so that

$$\sum_{1}^{N_1} \alpha_i x_i^{n_1} = \sum_{1}^{N_1} |x_i^{n_1}|.$$

Since (x_n) converges weakly to 0 we can (since $\ell_1^* = \ell_\infty$) choose n_{j_2} so large that $\sum\limits_{1}^{N_1} |x_i^{n_{j_2}}| < \frac{\epsilon}{5}$. and choose $N_2 > N_1$ so that $\sum\limits_{N_2+1}^{\infty} |x_i^{n_{j_2}}| < \frac{\epsilon}{5}$ and $\sum\limits_{1}^{N_2} |x_i^{n_{j_2}}| > \frac{4\epsilon}{5}$. Again choose $\alpha_{N_1+1} \cdots \alpha_{N_2}, |\alpha_i| = 1$, so that

$$\sum_{N_1+1}^{N_2} \alpha_i x_i^{n_{j_2}} = \sum_{N_1+1}^{N_2} |x_i^{n_{j_2}}| > \frac{4\epsilon}{5}.$$

Put the α_i together to form an element of ℓ_∞ and observe that the humps $\sum_{N_{k+1}}^{N_{k+1}} \alpha_i x_1^{n_{j_k}}$ are to big.

Analogous to the definition in VII, an operator $T \in \mathcal{L}(X,Y)$ is completely continuous if T maps weakly convergent sequences in X into sequences converging in the norm of Y.

3. Exercise 2 implies that the identity operator on ℓ_1 is completely continuous. Is it compact?

4. Show that in Lemma IX.4, (e_n) need not be complete, i.e. show that if $T \in K(H)$ and (e_n) is an orthonormal set in H then $\lim\limits_{n\to\infty} Te_n = 0$.

   Recall that it follows from IX.5 and IX.6 that if $T \in S_p(H)$ then $T \in K(H)$.

5. Show that if $T \in S_p(H)$ we can *factor* T as follows

$$
\begin{array}{ccc}
H & \xrightarrow{\;\;T\;\;} & H \\
A \downarrow & & \uparrow B \\
\ell_\infty & \xrightarrow{\;\;D\;\;} & \ell_p
\end{array}
$$

where D is the diagonal map given by

$$(\sigma_n(T)) \text{ i.e if } (a_n) \in \ell_\infty, D((a_n)) = (\sigma_n(T)a_n).$$

[HINT] Look at the Schmidt representation of T.

6. If $1 < p < \infty$ and $\frac{1}{p} + \frac{1}{q} = 1$ show that $\ell_p^* = \ell_q$ in the sense of Exercise 1.

   [HINT] In Exercise 5 in the notes to Chapter IV this result was given for $p = q = 2$. Mimic that proof replacing α_k (defined there) by

$$\alpha_k = \begin{cases} |\beta_k|^q sgn\, \bar{\beta}_k & \text{if } 1 \le k \le n \\ 0 & \text{if } k > n. \end{cases}$$

7. What is $\ell_p^{**} = \left(\ell_p^*\right)^*$?

X. HILBERT-SCHMIDT AND TRACE CLASS OPERATORS

Long before the introduction of the Schatten p-classes $\mathcal{S}_p$, a class of operators, the Hilbert-Schmidt operators, had been studied on Hilbert spaces.

X.1 Definition. We write $T \in HS(H)$, the Hilbert-Schmidt class, provided there is a complete orthonormal set (y_n) for H with
$$\sum_{n=1}^{\infty} \|Ty_n\|^2 < +\infty.$$

Example. Let (e_n) be the canonical complete orthonormal set for ℓ_2 and let $Tx = \sum_{n=1}^{\infty} \frac{1}{n}(x, e_n)e_n$. Clearly $\|Te_k\|^2 = \frac{1}{k^2}$ so $T \in HS(\ell_2)$.

X.2 Theorem. *Let $T \in \mathcal{L}(H)$. Then $T \in HS(H)$ if and only if*

$$T^* \in HS(H)$$

and if (x_n) is <u>any</u> *complete orthonormal set in H,*

$$\sum \|Tx_n\|^2 < +\infty.$$

All such sums have the same value.

Proof: Let (y_n) be a complete orthonormal set in H with $\sum \|Ty_n\|^2 < +\infty$. Let (x_n) be an arbitrary complete orthonormal set in H. Then by Parseval's equality $\|Ty_n\|^2 = \sum_{m=1}^{\infty} |(Ty_n, x_m)|^2$. Thus

$$\sum_{n=1}^{\infty} \|Ty_n\|^2 = \sum_{n=1}^{\infty} \sum_{m=1}^{\infty} |(Ty_n, x_m)|^2$$

$$= \sum_{m=1}^{\infty} \sum_{n=1}^{\infty} |(y_n, T^*x_m)|^2$$

$$= \sum_{m=1}^{\infty} \|T^*x_m\|^2$$

(the student should justify the interchange of summation and the last equality). Thus $\sum_{m=1}^{\infty} \|T^*x_m\|^2$ exists and is independent of the choice of complete orthonormal set (y_n). If $T^* \in HS(H)$ the above shows that

$T = T^{**} \in HS(H)$ and $\sum\limits_{n=1}^{\infty} \|Tx_n\|^2$ exists for every complete orthonormal set (x_n) in H and the sum is independent of the choice of the complete orthonormal set.

Thus we define the Hilbert-Schmidt norm $hs(T)$ by

$hs(T) = \left[\sum\limits_{n=1}^{\infty} \|Tx_n\|^2 \right]^{1/2}$ where $T \in HS(H)$ and (x_n) is a complete orthonormal set in H. Clearly Theorem X.2 shows that

$$hs(T) = hs(T^*).$$

Exercise 1.   a) Show that for $T \in \mathcal{L}(H)$, $\|T\| \leq hs(T)$.
[HINT] Choose, for $\varepsilon > 0$, an x_0, $\|x_0\| = 1$ with $\|T\|^2 < \|Tx_0\|^2 + \varepsilon$. Extend $\{x_0\}$ to complete orthonormal set for H.

b) Show that if $T \in HS(H)$ and (x_n) is a complete orthonormal set in H then $hs(T) = \sum\limits_{n=1}^{\infty} \sum\limits_{m=1}^{\infty} |(Tx_n, x_m)|^2$. Hint: Look at the proof of X.2.

X.3 Theorem.   *The Hilbert-Schmidt class is identical with the Schatten 2-class. That is, $HS(H) = \mathcal{S}_2(H)$ and $hs(T) = s_2(T) = \left[\sum\limits_{n=1}^{\infty} \sigma_n(T)^2 \right]^{1/2}$.*

Proof: Let $T \in HS(H)$, (x_n) a complete orthonormal set in H, $\varepsilon > 0$ and N such that

$$\left(\sum\limits_{n=N+1}^{\infty} \|Tx_n\|^2 \right)^{1/2} < \varepsilon.$$

Let $Px = \sum\limits_{i=1}^{N} (x, x_i)x_i$ be the orthogonal projection onto $[x_1, \ldots, x_n]$. Then for $x \in H$, $\|x\| \leq 1$,

$$\|(T - TP)x\| = \left\| \sum\limits_{n=N+1}^{\infty} (x, x_i)Tx_i \right\|$$

$$\leq \left(\sum\limits_{n=N+1}^{\infty} |(x, x_i)|^2 \right)^{1/2} \left(\sum\limits_{n=N+1}^{\infty} \|Tx_i\|^2 \right)^{1/2}$$

$$< \varepsilon$$

so $T \in K(H)$. Thus there are orthonormal sets (v_j), (u_j) with $Tx =$

$\sum\limits_{j=1}^{\infty} \sigma_j(T)(x, u_j)v_j$ (Schmidt representation). In particular,

$$\|Tx_n\|^2 = (Tx_n, Tx_n) = \sum_{j=1}^{\infty} \sigma_j(T)^2 |(x_n, u_j)|^2.$$

Thus

$$\sum_{n=1}^{\infty} \|Tx_n\|^2 = \sum_{n=1}^{\infty} \sum_{j=1}^{\infty} \sigma_j(T)^2 |(x_n, u_j)|^2$$

$$= \sum_{j=1}^{\infty} \sum_{n=1}^{\infty} \sigma_j(T)^2 |(x_n, u_j)|^2 = \sum_{j=1}^{\infty} \sigma_j(T)^2$$

i.e., $T \in \mathcal{S}_2(H)$ and $hs(T) = s_2(T)$. Now let $T \in \mathcal{S}_2(H)$ have Schmidt-representation $Tx = \sum\limits_{j=1}^{\infty} \sigma_j(T)(x, u_j)v_j$ and let (x_n) be a complete orthonormal set in H. Retrace the steps above obtaining $T \in HS(H)$ and $hs(T) = s_2(T)$.

Thus we will drop the artificial HS, hs and use $\mathcal{S}_2, s_2$ for the Hilbert-Schmidt operators and the Hilbert-Schmidt norm.

Exercise 2. Let (x_n) be a complete orthonormal set in H and for $S, T \in \mathcal{S}_2(H)$ define $(S, T) = \sum\limits_{n}(Tx_n, Sx_n)$. Show that (T, S) is an inner-product on $\mathcal{S}_2(H)$ and $(T, T)^{1/2} = s_2(T)$. In particular, $\mathcal{S}_2(H)$ is a Hilbert space. (I know of no use for this representation of Hilbert space other than it's interesting!)

To motivate the *final* class of operators we will study, and understand the title of the monograph, we need to recall some concepts from linear algebra. The student who has forgotten this material should spend an hour reviewing any decent linear algebra book.

Thus let $(x_1, \ldots, x_n)$ be a basis for an n-dimension complex Hilbert space E_n. An operator $A \in \mathcal{L}(E_n)$ is completely determined by its values at x_i, $i = 1, \ldots, n$. Thus, let $Ax_i = \sum\limits_{j=1}^{n} a_{ij}x_j$. Clearly A corresponds in a natural way to the matrix $\widetilde{A} = (a_{ij})$. We have used this before! The *trace* of A, tr$A$, is defined by tr$A = \sum\limits_{i=1}^{n} a_{ii}$, the sum of the diagonal elements of

$\widetilde{A}$. If $A, B \in \mathcal{L}(E_n)$, $\mathrm{tr}(AB) = \mathrm{tr}(BA)$. Indeed if $Bx_i = \sum\limits_{j=1}^{n} b_{ij}x_j$ then

$$ABx_i = \sum_{j=1}^{n}\sum_{k=1}^{n} b_{ij}a_{jk}x_k$$

and

$$BAx_i = \sum_{j=1}^{n}\sum_{k=1}^{n} a_{ij}b_{jk}x_k$$

so

$$\mathrm{tr}(AB) = \sum_{i=1}^{n}\sum_{j=1}^{n} b_{ij}a_{ji} = \sum_{i=1}^{n}\sum_{j=1}^{n} a_{ji}b_{ij} = \mathrm{tr}(BA).$$

Also, $\mathrm{tr}A$ is independent of the choice of basis: if $\{y, \dots, y_n\}$ is another basis for E_n let C be the basis transformation $Cx_i = y_i$. Then $C, C^{-1} \in \mathcal{L}(E_n)$ and

$$CAC^{-1}y_i = \sum_{j=1}^{n} a_{ij}y_i \qquad (\text{where } Ax_i = \sum_{j=1}^{n} a_{ij}x_j)$$

so by what was just shown $\mathrm{tr}A$ (with respect to (y_i)) $= \mathrm{tr}(C^{-1}CA) = \mathrm{tr}(CAC^{-1}) = \sum\limits_{i=1}^{n} a_{ii} = \mathrm{tr}A$ (with respect to (x_i)).

Superficially $\mathrm{tr}A$ appears to have little to do with the eigenvalues of A. However the characteristic polynomial $\det(\lambda I - \widetilde{A})$ of A (I the identity on E_n) is a polynomial of the form $\lambda^n - c_1\lambda^{n-1} + \cdots$ whose roots are the eigenvalues of A. For any monic polynomial $p(x) = x^n + a_1x^{n-1} + \cdots$ the sum of the roots of $p(x)$ is just $-a_1$. Thus if $\lambda_1, \dots, \lambda_n$ are the eigenvalues of A (counting multiplicities) we obtain $c_1 = \sum\limits_{i=1}^{n} \lambda_i$. On the other hand, from the definition of the determinant $c_1 = \sum\limits_{i=1}^{n} a_{ii}$, i.e., $\mathrm{tr}A = \sum\limits_{i=1}^{n} \lambda_i$, the sum of the eigenvalues of A. Since $a_{ij} = (Tx_i, x_j)$ we also have $\mathrm{tr}A = \sum\limits_{i=1}^{n}(Tx_i, x_i)$.

Thus, in the middle 30's, von-Neumann and Murray were set to the problem of finding operators on H (now infinite dimensional) having a "trace".

Analogous to the above, for T in "some class of operators on H" a "fake trace" should be of the form: $\sum\limits_{n=1}^{\infty}(Tx_i, x_i)$ where (x_i) is a complete orthonormal set in H. The example $Tx = \sum \frac{1}{n}(x, e_n)e_n \in \mathcal{S}_2(\ell_2)$ shows that $\mathcal{S}_2$, the Hilbert-Schmidt operators, does not support a trace in this sense.

X.4 Definition.   A $T \in \mathcal{L}(H)$ is in the *trace class* (written temporarily $T \in TC(H)$) provided $Tx = \sum_{n=1}^{\infty}(x, y_n)x_n$ where $\sum \|x_n\| \, \|y_n\| < +\infty$.

Let $\nu(T) = \inf \sum \|y_n\| \, \|x_n\|$ where the infimum is taken over all such representations of $T \in TC(H)$.

X.5 Theorem.   *The trace class is identical with the Schatten 1-class. That is, $TC(H) = \mathcal{S}_1(H)$ and $s_1(T) = \nu(T)$ for $T \in TC(H)$.*

Proof: Clearly, if $T \in TC(H)$ then T is the limit of finite rank operators in the operator norm so T has a Schmidt representation

$$Tx = \sum_{n=1}^{\infty} \sigma_n(T)(x, x_i)y_i, \quad (x_i), (y_i) \text{ orthonormal sets }.$$

For $\varepsilon > 0$ choose a representation of T, $Tx = \sum_{n=1}^{\infty}(x, w_n)z_n$, where

$$\sum_{n=1}^{\infty} \|w_n\| \, \|z_n\| < (1 + \varepsilon)\nu(T).$$

Then

$$\sum_{n=1}^{\infty} \sigma_n(T) = \sum_{n=1}^{\infty}(Tx_n, y_n) = \sum_{n=1}^{\infty} \sum_{m=1}^{\infty}(x_n, w_m)(z_m, y_n)$$

$$\leq \sum_{m=1}^{\infty} \left(\sum_{n=1}^{\infty} |(x_n, w_m)|^2 \right)^{1/2} \left(\sum_{n=1}^{\infty} |(z_m, y_n)|^2 \right)^{1/2}$$

$$\leq \sum_{m=1}^{\infty} \|w_m\| \, \|z_m\| < (1 + \varepsilon)\nu(T)$$

(Hölder and Bessel inequalities), i.e., $s_1(T) \leq \nu(T) < +\infty$ so $T \in \mathcal{S}_1(H)$. Clearly, if $T \in \mathcal{S}_1(H)$ has a Schmidt representation

$$Tx = \sum_{n=1}^{\infty} \sigma_n(T)(x, z_n)x_n$$

then

$$Tx = \sum_{n=1}^{\infty}(x, y_n)x_n, \quad y_n = \sigma_n(T)z_n$$

and

$$\sum_{n=1}^{\infty}\|y_n\|\,\|x_n\| = \sum_{n=1}^{\infty}\sigma_n(T)$$

so $T \in TC(H)$ and $\nu(T) \leq s_1(T)$.

Exercise 3.   (i) Let $(\lambda_n) \in \ell_1$. Let $\frac{1}{p} + \frac{1}{q} = 1$, $1 < p < +\infty$. Show that there are $(\alpha_n) \in \ell_p$, $(\beta_n) \in \ell_q$ with $\lambda_n = \alpha_n\beta_n$ for all n. If $p = 1$ or ∞ show that $\lambda_n = \alpha_n\beta_n$, $(\alpha_n) \in c_0$, $(\beta_n) \in \ell_1$.

(ii) Show that $\mathcal{S}_1(H) = \mathcal{S}_p(H) \circ \mathcal{S}_q(H)$ (i.e., each $T \in \mathcal{S}_1(H)$ is of the form $T = AB$ where $B \in \mathcal{S}_q$, $A \in \mathcal{S}_p$, $\frac{1}{p} + \frac{1}{q} = 1$, ($B \in K(H)$, $A \in \mathcal{S}_1(H)$ in case $p = 1$ or ∞), and each $T = AB$, $A \in \mathcal{S}_q$, $B \in \mathcal{S}_p$ is in $\mathcal{S}_1$. Moreover show that $s_1(T) \leq s_p(A)s_q(B)$ when one has such a composition.

(iii) We now (obviously) have $\mathcal{S}_1(H) \subset \mathcal{S}_2(H) \subset K(H)$. We know also that for $T \in K(H)$, $\sigma(T) = \{\lambda_n\} \cup \{0\}$ where $(\lambda_n) \in c_0$. Show that for $T \in \mathcal{S}_p(H)$, $(\lambda_n) \in \ell_p$ and in particular, for $T \in \mathcal{S}_1(H)$, $(\lambda_n) \in \ell_1$.

X.6 Theorem.   *Let $S, T \in \mathcal{S}_2(H)$ and (x_n) be a complete orthonormal set in H. Then $\sum_{n=1}^{\infty}|(Sx_n, T^*x_n)| < +\infty$ and $\sum(Sx_n, T^*x_n)$ is independent of the choice of complete orthonormal set.*

Proof: Let (y_n) be a complete orthonormal set in H. Then

$$\sum_{n=1}^{\infty}\sum_{m=1}^{\infty}|(Sx_n, y_m)\overline{(T^*x_n, y_m)}| \leq \sum_{n=1}^{\infty}\left(\sum_{m=1}^{\infty}|(Sx_n, y_m)|^2\right)^{1/2}$$

$$\left(\sum_{m=1}^{\infty}|(T^*x_n, y_m)|^2\right)^{1/2} = \sum_{n=1}^{\infty}\|Sx_n\|\,\|T^*x_n\| \leq \left(\sum_{n=1}^{\infty}\|Sx_n\|^2\right)^{1/2}$$

$$\left(\sum_{n=1}^{\infty}\|T^*x_n\|^2\right)^{1/2} = s_2(S)s_2(T^*).$$

(What inequalities are being used?) Since this series is absolutely convergent, we may switch the order of summation and so

$$\sum_{n=1}^{\infty}(Sx_n, T^*x_n) = \sum_{n=1}^{\infty}\sum_{m=1}^{\infty}(Sx_n, y_m)\overline{(T^*x_n, y_m)}$$

$$= \sum_{m=1}^{\infty}\sum_{n=1}^{\infty}(Ty_m, x_n)\overline{(S^*y_m, x_n)}.$$

(why?) Similarly, $\sum_{m=1}^{\infty}(Sy_m, T^*y_m)$ has this same value.

X.7 Definition. Let $A, B \in S_2(H)$ and (x_n) be a complete orthonormal set in H. Let

$$\mathrm{tr}(A, B) = \sum_{n=1}^{\infty}(Ax_n, B^*x_n).$$

Observe that if $T \in S_1(H)$ then $T = B \circ A$ for some $A, B \in S_2(H)$ (exercise 3). This factorization, of course, if far from unique. However, if $T = D \circ C$, $C, D \in S_2(H)$ then

$$\mathrm{tr}(A, B) = \sum_{n=1}^{\infty}(Ax_n, B^*x_n) = \sum_{n=1}^{\infty}(BAx_n, x_n)$$

$$= \sum_{n=1}^{\infty}(Tx_n, x_n) = \sum_{n=1}^{\infty}(DCx_n, x_n)$$

$$= \sum_{n=1}^{\infty}(Cx_n, D^*x_n) = \mathrm{tr}(C, D).$$

Thus we are let to the following:

X.8 Definition. For $T \in S_1(H)$ we define the *functional trace* by $\phi - \mathrm{tr}(T) = \sum_{n=1}^{\infty}(Tx_n, x_n)$ where (x_n) is a complete orthonormal set in H.

By exercise 3 and the above remark $\phi - \mathrm{tr}(T)$ is independent of the choice of complete orthonormal set (x_n).

Our next result shows that $\phi - \mathrm{tr}(T)$ is given as a sum of "diagonal elements" analogous to the finite dimensional case.

X.9 Theorem. *Let $T \in S_1(H)$ have a representation $Tx = \sum_{n=1}^{\infty}(x, y_n)z_n$ where $\sum \|y_n\| \|z_n\| < +\infty$. Then*

$$\phi - \mathrm{tr}\, T = \sum_{n=1}^{\infty}(z_n, y_n).$$

Proof: Let (x_m) be a complete orthonormal set in H. Then

$$\sum_{m=1}^{\infty}(Tx_m, x_m) = \sum_{m=1}^{\infty}\sum_{n=1}^{\infty}(x_m, y_n)(z_n, x_m)$$

$$= \sum_{n=1}^{\infty}\left(\sum_{m=1}^{\infty}(x_m, y_n)x_m, \sum_{\ell=1}^{\infty}(x_\ell, z_n)x_\ell\right)$$

$$= \sum_{n=1}^{\infty}\left(\sum_{m=1}^{\infty}\overline{(y_n, x_m)}x_m, \sum_{\ell=1}^{\infty}\overline{(z_n, x_\ell)}x_\ell\right)$$

$$= \overline{\sum_{n=1}^{\infty}\left(\sum_{m=1}^{\infty}(y_n, x_m)x_m, \sum_{\ell=1}^{\infty}(z_n, x_\ell)x_\ell\right)}$$

$$= \sum_{n=1}^{\infty}\overline{(y_n, z_n)} = \sum_{n=1}^{\infty}(z_n, y_n).$$

Of course, the student should check these computations.

X.9 Corollary. $\phi - \text{tr}(\bullet)$ is continuous with respect to the S_1-norm; that is, $\phi - \text{tr}(\bullet)$ is a continuous linear functional on S_1.

Proof: If $Tx = \sum(x, y_n)z_n$, $Sx = \sum(x, u_n)v_n$ are in S_1 and

$$\sum_{n=1}^{\infty}\|y_n\|\,\|z_n\| < +\infty, \quad \sum_{n=1}^{\infty}\|u_n\|\,\|v_n\| < +\infty$$

then $(S + T)x = \sum(x, s_n)r_n$ where $s_{2n} = y_n$, $s_{2n-1} = u_n$, $r_{2n} = z_n$, $r_{2n-1} = v_n$ (for example). Then, from the above,

$$\phi - \text{tr}(S + T) = \sum(r_n, s_n) = \phi - \text{tr}(S) + \phi - \text{tr}(T).$$

Also $|\phi - \text{tr } T| = |\sum_{n=1}^{\infty}(z_n, y_n)| \leq \sum_{n=1}^{\infty}\|z_n\|\,\|y_n\|$ and since $Tx = \sum(x, y_n)z_n$ is an arbitrary representation of T with $\sum\|y_n\|\,\|z_n\| < +\infty$ it follows from X.5 that $|\phi - \text{tr } T| \leq s_1(T)$.

Two big questions remain. From the weak Weyl inequality it follows that for $T \in S_1(H)$, the eigenvalues $(\lambda_n(T)) \in \ell_1$ and satisfy $\sum_{n=1}^{\infty}|\lambda_n(T)| \leq s_1(T)$. Let us define on $S_1(H)$ a "spectral trace"

$$\sigma - \text{tr } T = \sum_{n=1}^{\infty}\lambda_n(T).$$

Is $\sigma - \mathrm{tr}(\bullet)$ linear and continuous on $\mathcal{S}_1(H)$? Is $\sigma - \mathrm{tr}(\bullet) = \phi - \mathrm{tr}(\bullet)$ on $\mathcal{S}_1(H)$ (as in the finite dimensional case)? Neither question is easy to answer and the remaining chapter is devoted entirely to these questions.

Remarks, Exercises and Hints

Many of the classes of operators on Hilbert space are defined in terms unique to these spaces. Part of the fun of studying operator theory on Banach spaces is to formulate concepts which make sense in arbitrary Banach spaces and which, when applied to Hilbert space, are equivalent to the original definitions.

As we know if (x_n) is a complete orthonormal set in H, $\sum_{n=1}^{\infty} |(x_n, x)|^2 < +\infty$ for all $x \in H$. By the Riesz Representation theorem this is equivalent to saying $\sum_{n=1}^{\infty} |f(x_n)|^2 < +\infty$ for each $f \in H^*$.

We say that a sequence (y_n) in a Banach space X is *weakly square summable* provided

$$\sum_{n=1}^{\infty} |f(y_n)|^2 < +\infty$$

for each $f \in X^*$; and, *square summable* provided

$$\sum_{n=1}^{\infty} ||y_n||^2 < +\infty.$$

An operator $T \in \mathcal{L}(X, Y)$ is *absolutely 2-summing* provided T maps weakly square summable sequences in X into square summable sequences in Y.

This notion of absolutely 2-summing operators (named very differently) is due to A. Grothendieck, probably motivated by earlier (1933) work of W. Orlicz. We write $\prod_2(X, Y)$ for the absolutely 2-suming operators and say T is a $\prod_2$ operator. You probably have guessed by now that this is a correct generalization to Banach spaces of the Hilbert-Schmidt operators.

A set A in a Banach space X is *weakly bounded* provided $\sup\{|f(a)| : a \in A\} < +\infty$ for each $f \in X^*$.

1. Using the hint in problem 1 Chapter VII Notes, Exercises and Hints show that if A is weakly bounded then A is bounded in X, i.e.,

$$\sup\{||a|| : a \in A\} < +\infty.$$

2 Suppose (y_n) is a weakly square summable sequence in a Banach space X. Show that

$$\epsilon_2(y_n) = \sup\left\{ \left(\sum_{n=1}^{\infty} |f(y_n)|^2 \right)^{\frac{1}{2}} : ||f|| \leq 1 \right\} < +\infty.$$

[HINT] Let

$$A = \left\{ \sum_{n=1}^{\infty} \alpha_n y_n : \sum_{n=1}^{\infty} |\alpha_n|^2 \leq 1 \right\}.$$

Show that A is weakly bounded (use Holder's inequality) and then use 1 to show that A is bounded. Then for $f \in X^*, ||f|| \leq 1$ there is

$$(\alpha_i) \in \ell_2, \Sigma \alpha_i^2 \leq 1$$

with

$$\sum_{i=1}^{\infty} \alpha_i f(y_i) = \left(\Sigma |f(y_i|^2) \right)^{\frac{1}{2}} \text{ (why ?)} .$$

3. Show that if H is a Hilbert space $\prod_2(H) = S_2(H)$.

[HINT] That $\prod_2(H) \subset S_2(H)$ is obvious. If $T \in S_2(H)$ and (y_n) is a weakly square summable sequence in H choose a complete orthonormal set (x_i) for H and compute:

$$\sum_{i=1}^{N} ||Ty_i|| = \sum_{i=1}^{N} \left(\sum_{m=1}^{\infty} |(Ty_i, x_m)| \right)^2$$

$$= \sum_{m=1}^{\infty} \sum_{i=1}^{N} \frac{|(y_i, T^*x_m)|^2}{||T^*x_m||^2} ||T^*x_m||^2.$$

To aid in proofs of the exercises below, it is convenient to have a *finite* formulation of the definition of $\prod_2$ operators. If X is a Banach space and $x_1 \cdots x_n \in X$ let

$$\epsilon_2((x_i)) = \sup\left\{ \left(\sum_{i=1}^{n} |f(x_i)|^2 \right)^{\frac{1}{2}} : f \in X^*, ||f|| = 1 \right\}$$

and

$$\alpha_2((x_i)) = \left(\sum_{i=1}^{n} ||x_i||^2 \right)^{\frac{1}{2}} .$$

4. Show that $T \in \mathcal{L}(X, Y)$ is absolutely 2-suming if and only if there is a constant K (depending only on X and Y) such that $\alpha_2(T(x_i)) \leq K\epsilon_2((x_i))$ for all $x_1 \cdots x_n \in X$.

[HINT] If such a K exists use in conjunction with problem 2 to show that if (y_n) is weakly square summable, $\sum ||Ty_n||^2$ is Cauchy. If no such K exist, construct a weakly square summable square (y_n) with

$$\sum_{n=1}^{\infty} ||Ty_n||^2 = +\infty.$$

5. If one defines, for $T \in \prod_2(X,Y), \pi_2(T) = \inf K$, where K satisfies $\alpha_2\big(T(x_i)\big) \leq K\epsilon_2\big((x_i)\big)$ then $\pi_2(T)$ is a norm.

6. In 3 show that $s_2(T) = \pi_2(T)$.

The $\prod_2$ operators and more generally the $\prod_p$-operators $1 \leq p < +\infty$ (guess the definitions) have played a fundamental role in the structure theory of Banach spaces.

Grothendieck and, independently, A. F. Ruston also noticed that the definition of nuclear operator makes sense in an arbitrary Banach space. An operator $T \in \mathcal{L}(X,Y)$ is *nuclear* provided there are (f_n) in $X^*, (y_n)$ in Y such that

$$Tx = \sum_{n=1}^{\infty} f_n(x)y_n$$

and

$$\sum_{n=1}^{\infty} ||f_n||\,||y_n|| < +\infty.$$

We write $T \in N(X,Y)$ if T is nuclear.

7. Show that if $T \in N(X,Y)$ there are (λ_n) in $\ell_1, (g_n) \in X^*, with$ $\lim_{n \to \infty} ||g_n|| = 0$ and (z_n) in Y, with $\sup_n ||z_n|| < +\infty.$ such that

$$Tx = \sum_{n=1}^{\infty} \lambda_n g_n(x)z_n.$$

[Recall X, Exercise 3].

8. Use 7 to deduce Grothendieck's factorization theorem for nuclear operators: if

$$T \in N(X,Y),$$

$$
\begin{array}{ccc}
X & \xrightarrow{\ \ T\ \ } & Y \\
\downarrow{\scriptstyle A} & & \uparrow{\scriptstyle B} \\
c_0 & \xrightarrow{\ \ D\ \ } & \ell_1
\end{array}
$$

where D is a diagonal mapping $D((\alpha_n)) = (\lambda_n \alpha_n)$, $(\alpha_n) \epsilon c_o$ and (λ_n) as in 7.

Grothendieck also observed that, in general, the eigenvalues of $T \in N(X)$ are square summable and that this is the best possible result. This shows how difficult things are when one gets away from Hilbert space. Indeed the eigenvalues $(\lambda_n(T))$ of $T \in N(X)$ satisfy $\sum_{n=1}^{\infty} |\lambda_n(T)| < +\infty$ for all such T if and only if X is isomorphic to a Hilbert space. It is now also known that any sequence (λ_n) in ℓ_2, $\lambda_n \neq 0$, is the eigenvalue sequence of some nuclear operator. Things really go awry in Banach spaces!

If $\mathcal{R}(A)$ is contained in the domain of B and $\mathcal{R}(B)$ is contained in the domain of A then the operator pair (A, B) are called related operators:

$$X \xrightarrow{A} Y \xrightarrow{B} X.$$

Here X and Y are Banach spaces.

9. If (A, B) are related operators, show that AB and BA have the same eigenvalues.

There is a famous theorem concerning the factorization of $T \in \prod_2(X, Y)$:

Grothendieck-Pietsch Theorem: *If $T \in \prod_2(X, Y)$ there is a Hilbert space H and a compact space M such that*

$$
\begin{array}{ccc}
X & \xrightarrow{\ \ T\ \ } & Y \\
\downarrow{\scriptstyle A} & & \uparrow{\scriptstyle B} \\
C(M)) & \xrightarrow{\ \ J\ \ } & H
\end{array}
$$

and $J \in \prod_2(C(M), H)$. Here $C(M)$ denotes the continuous functions on M.

10. Prove the following easy version of the Grothendick-Pietsch Theorem: Show that if $T \in S_2(H)$ then T admits a factorization

$$
\begin{array}{ccc}
H & \xrightarrow{\;\;T\;\;} & H \\
A \downarrow & & \uparrow B \\
\ell_\infty & \xrightarrow{\;\;D\;\;} & \ell_2
\end{array}
$$

where D is the diagonal mapping

$$D\big((a_n)\big) = (\sigma_n(T)a_n), (a_n) \in \ell_\infty.$$

The space ℓ_∞ is $C(M)$ for suitable M but that result takes us too far afield!

11. Show that if $T \in \prod_2(X,Y)$, $S \in \mathcal{L}(Z,X)$, and $R \in \mathcal{L}(Y,W)$ then $RTS \in \prod_2(Z,W)$.

12. Use, 3, 8, 9, 10 and the Grothendieck-Pietsch factorization theorem to show that if $T \in \prod_2(X)$ then the eigenvalues of T, $\big(\lambda_n(T)\big)$, satisfy

$$\sum_{n=1}^{\infty} |\lambda_n(T)|^2 < +\infty.$$

13. Show that $N(X) \subset \prod_2(X)$ and deduce Grothendieck's result concerning the summability of the eigenvalues of $T \in N(X)$.

There is still other routes one can attempt to generalize trace class operators. The square root of positive self adjoint operators only makes sense on Hilbert space and thus the singular numbers $\big(\sigma_n(T)\big)$ defined as the eigenvalues of $\big(T^*T\big)^{\frac{1}{2}}$ also only makes sense on these spaces. But, the characterization given by VIII Exercise 4

$$\sigma_n(T) = \inf\big\{\|T - A\| : \text{ rank } A < n\big\}$$

makes sense in arbitrary Banach spaces. This led Pietsch to define the approximation numbers of $T \in \mathcal{L}(X,Y)$ by

$$\alpha_n(T) = \inf\big\{\|T - A\| : A \in \mathcal{L}(X,Y), \text{ rank } A < n\big\}$$

and to define an operator T of type ℓ_1 by requiring $\sum_{n=1}^{\infty} \alpha_n(T) < +\infty$. We write, in this case, $T \in \ell_1(X,Y)$.

14. Show that the approximation numbers $\big(\alpha_n(T)\big)$ satisfy the properties listed in Exercise 1 in the notes to Chapter VIII. In particular, show $\alpha_n(ST) \leq \|S\|\alpha_n(T)$.

15. Let $T \in \mathfrak{L}(\ell_2)$. Let T_n be the restriction of T to $\mathrm{sp}[e, \cdots e_n] = \ell_2^n$. Then $\alpha_i(T_n) \leq \alpha_i(T)$ for all i.

[HINT] Factor

$$
\begin{array}{ccc}
\ell_2^n & \xrightarrow{\;\;T_n\;\;} & \ell_2^n \\
\Big\downarrow {\scriptstyle J} & & \Big\uparrow {\scriptstyle P} \\
\ell_2 & \xrightarrow{\;\;T\;\;} & \ell_2
\end{array}
$$

where J is the inclusion map and P the orthogonormal projection onto ℓ_2^n.

16. Let I be the identity an ℓ_2. Show that $\alpha_i(I) = 1$ for each i

    [HINT] For one inequality let P be a suitable projection of rank less than n. For the other, if the rank of $A \neq 0$ is less than n, choose $x \epsilon \ker A$ with $\|x\| = 1$].

    Let $D : \ell_2 \to \ell_2$ be a diagonal map given by (λ_i) where $\lambda_1 \geq \lambda_2 \geq \cdots > 0$. That is, $D(\beta) = (\lambda_i \beta_i)$ for each $\beta = (\beta_i) \in \ell_2$.

17. For such a D show that $\alpha_n(D) = \lambda_n$.

    [HINT] Clearly $D(\beta) = \sum_{i=1}^{\infty} \lambda_i \beta_i e_i$ where (e_i) is the canonical orthonormal set for ℓ_2. Let A be the truncated operator

    $$
    A(\beta) = \sum_{i=1}^{n-1} \lambda_i \beta_i e_i.
    $$

Then

$$
\|(D - A)\beta\| = \lambda_n \left(\sum_{i=1}^{\infty} \left(\frac{\lambda_i}{\lambda_n} \beta_i \right)^2 \right)^{\frac{1}{2}}.
$$

For the other inequality observe that the identity operator I_n on ℓ_2^n factors

$$
\begin{array}{ccc}
\ell_2^n & \xrightarrow{\;\;I_n\;\;} & \ell_2^n \\
{\scriptstyle D_1} \searrow & & \nearrow {\scriptstyle D_n} \\
& \ell_2^n &
\end{array}
$$

where D_1 is the diagonal corresponding to $\left(\frac{1}{\lambda_i} \right)_{i=1}^{n}$ and D_n is the restriction of D to ℓ_2^n. Thus

$$
1 = \alpha_k(I_n) = \alpha_k(D_n D_1).]
$$

We can now give a factorization argument to show that for $T \in K(H), \sigma_n(T) = \alpha_n(T)$.

(See Exercise 6, Chapter VIII)

18. Let $T \in K(H)$. Show that

$$
\begin{array}{ccc}
H & \xrightarrow{\quad T \quad} & H \\
A \Big\Updownarrow A^* & & B^* \Big\Updownarrow B \\
\ell_2 & \xrightarrow{\quad D \quad} & \ell_2
\end{array}
$$

where D is a diagonal operator. [HINT] Let $Th = \Sigma \sigma_n(T)(h, x_n)y_n$ be a Schmidt decomposition of T. Let

$$A(h) = \sum_{n=1}^{\infty}(h, x_n)e_j, \quad B((\beta)) = \sum_{n=1}^{\infty}(\beta_n, e_n)y_n$$

and

$$D(\beta_n) = (\sigma_n(T)\beta_n)$$

for $\beta = (\beta_n) \in \ell_2$. Then show that

$$T = BDA \text{ and } D = B^* TA^*,$$

where A^*, B^* are the Banach space adjoints and e_n is the canonical orthonormal set for ℓ_2. Compute

$$A^*(e_n) \text{ and } B^*(y_n).)$$

19. (Allakhverdief): For $T \in K(H), \sigma_n(T) = \alpha_n(T)$.
    [HINT] Use 16, 17, and 18 for a "factorization" proof.

The eigenvalues of $T \in \ell_1(X)$ satisify $\sum_{n=1}^{\infty} |\lambda_n(T)| < +\infty$ for each Banach space X. Observe that if H is a Hilbert space $N(H) = \ell_1(H)$. since $\alpha_n(T) = \sigma_n(T)$.

Curiously, $N(X) = \ell_1(X)$ if and only if X is isomorphic to a Hilbert space.

To paraphrase Barry Simons, operator theory on Banach spaces is a zoo. But, zoos are fun to visit!

XI. THE LIDSKIJ TRACE THEOREM

The purpose of this last chapter is to show that for $T \in S_1(H)$

$$\phi - \text{tr } T = \sigma - \text{tr } T.$$

This result is where all the preceding material has been heading. It was proved by Lidskij in 1957 — more than twenty years after the trace class operators were introduced! Grothendieck *said* it was true in 1955!

There are now several proofs of this result in the literature. All are quite difficult. One of the difficulties is that it is possible for $T \in S_1(H)$, $T \neq 0$ and $T^2 = 0$. For such a T, $\sigma(T) = \{0\}$. Why should $\phi - \text{tr } T = 0$?

On the other hand, even though it holds in the finite dimensional case, there is no real reason to expect $\sigma - \text{tr}(T)$ to be linear or continuous on $S_1(H)$. Thus, the Lidskij result is quite remarkable.

Our proof follows that of Leiterer and Pietsch. The main ideas are due to König.

We begin by showing that for $S, T \in S_1(H)$ with eigenvalues $(\lambda_i(S)), (\lambda_i(T))$ respectively,

$$\sum_{i=1}^{\infty} \lambda_i(S + T) = \sum_{i=1}^{\infty} \lambda_i(S) + \sum_{i=1}^{\infty} \lambda_i(T),$$

i.e., σ trace is linear. We begin with finite rank operators.

XI.1 Theorem. *Let T, S have finite dimensional ranges. Then*

$$\sigma - \text{tr}(S + T) = \sigma - \text{tr}(S) + \sigma - \text{tr}(T).$$

Proof: We can assume that S and T have the same number of terms in forming $\sigma - \text{tr}(\bullet)$ (put in some zeros!) Recall that $\mathcal{R}(T)$ denotes the range of T. Let

$$M = \text{span}[\mathcal{R}(T) \cup \mathcal{R}(S) \cup \mathcal{R}(S + T)].$$

Then M is finite dimensional and if U is any of the operators T, S or $T + S$, $U(M) \subset M$. Let T_M, etc. denote the restriction to M. Then these operators may be viewed as matrices and so by our previous discussion in chapter VIII,

$$\sigma - \text{tr}(T_M + S_M) = \sigma - \text{tr}(T_M) + \sigma - \text{tr}(S_M).$$

But T_M and S_M have the same eigenvalues as T and S respectively. This proves the result.

The next part is standard: If f is defined on a dense subspace of a Banach space X and is uniformly continuous, then f has a *unique continuous* extension to all of X. Indeed, we used this fact in the proof of the polar decomposition theorem (VIII.5).

Exercise 1. Prove this remark.

Now let $D = \{T \in \mathcal{L}(H) \,|\, T \text{ has finite rank}\}$. Clearly D is dense in $S_1(H)$ with respect to the S_1-norm. By the weak Weyl inequality and XI.1 we have for $S, T \in D$

$$|\sigma - \mathrm{tr}(S) - \sigma - \mathrm{tr}(T)| = |\sigma - \mathrm{tr}(S-T)| = |\sum_{i=1}^{\infty} \lambda_i(S-T)|$$

$$\leq \sum_{i=1}^{\infty} |\lambda_i(S-T)| \leq s_1(S-T),$$

i.e., $\sigma - \mathrm{tr}(\bullet)$ is linear and continuous (with respect to the S_1-norm) on D and so has a unique extension, $\widetilde{\mathrm{tr}}$ to all of $S_1(H)$.

It appears, we have made things worse. We now have three candidates for traces on $S_1(H)$: $\sigma - \mathrm{tr}$, $\phi - \mathrm{tr}$ and $\widetilde{\mathrm{tr}}$. We can however immediately get back to our original two trace functionals.

XI.2 Theorem. If $T \in S_1(H)$, $\widetilde{\mathrm{tr}}(T) = \phi - \mathrm{tr}\, T$.

Proof: Suppose first that T has rank 1. Then $Tx = (x,v)u$ for some $u, v \in H$ and $\phi - \mathrm{tr}\, T = (u,v)$. If $Tx = \lambda x$, $\lambda \neq 0$, $x \neq 0$ then $(x,v)u = \lambda x$ so $u = \frac{\lambda}{(x,v)}x$. Thus

$$\sigma - \mathrm{tr}(T) = (u,v) = (\frac{\lambda}{(x,v)}x, v) = \lambda.$$

Now suppose rank $T = n$, so $Tx = \sum_{i=1}^{n}(x,v_i)u_i$ for suitable v_i, u_i in H. By linearity and what was just shown

$$\sigma - \mathrm{tr}\, T = \sum_{i=1}^{n} \sigma - \mathrm{tr}\,([(\bullet, v_i)u_i]) = \sum_{i=1}^{n}(u_i, v_i) = \phi - \mathrm{tr}\, T.$$

Thus $\widetilde{\mathrm{tr}}(T) = \phi - \mathrm{tr}\, T$ if rank $T = n$. But $\phi - \mathrm{tr}(\bullet)$ is linear and continuous on $\mathcal{S}_1(H)$ and agrees with $\widetilde{\mathrm{tr}}$ on a dense set. By uniqueness $\widetilde{\mathrm{tr}}(T) = \phi - \mathrm{tr}(T)$ if $T \in \mathcal{S}_1(H)$.

We are still a long way from showing that $\sigma - \mathrm{tr}(T) = \phi - \mathrm{tr}(T)$ for $T \in \mathcal{S}_1(H)$.

A next step in the proof is yet another inequality.

XI.3 Theorem. (Hardy's inequality). Let $p > 1$, m a positive integer, $a_i \geq 0$ and $A_n = \sum\limits_{i=1}^{n} a_i$. Then

$$\sum_{n=1}^{m} \left(\frac{A_n}{n}\right)^p \leq \left(\frac{p}{p-1}\right)^p \sum_{n=1}^{m} a_n^p.$$

Proof: There is no loss of generality in supposing $a_i > 0$. [If $a_1 = 0$ writing $b_i = a_{i+1}$ one obtains the weaker inequality $\left(\frac{b_1}{2}\right)^p + \left(\frac{b_1+b_2}{3}\right)^p \cdots \left(\frac{p}{p-1}\right)^p \sum\limits_{i=1}^{m} b_i^p.$] Let $\frac{1}{p} + \frac{1}{q} = 1$. Then

$$\left(\frac{A_n}{n}\right)^p - \frac{p}{p-1}\left(\frac{A_n}{n}\right)^{p-1} a_n$$

$$= \left(\frac{A_n}{n}\right)^p - \frac{p}{p-1}\left\{n\frac{A_n}{n} - (n-1)\frac{A_{n-1}}{n-1}\right\}\left(\frac{A_n}{n}\right)^{p-1}$$

$$= \left(\frac{A_n}{n}\right)^p \left(1 - \frac{np}{p-1}\right) + \frac{(n-1)p}{p-1}\left(\frac{A_n}{n}\right)^{p-1}\frac{A_{n-1}}{n-1}$$

$$\leq \left(\frac{A_n}{n}\right)^p \left(1 - \frac{np}{p-1}\right) + \frac{(n-1)p}{p-1}\left[\frac{\left(\frac{A_n}{n}\right)^{(p-1)q}}{q} + \frac{\left(\frac{A_{n-1}}{n-1}\right)^p}{p}\right]$$

$$= \frac{p-1}{p-1}\left(\frac{A_n}{n}\right)^p \left[\left(1 - \frac{np}{p-1}\right)\right] +$$

$$\frac{(n-1)}{p-1}\left[(p-1)\left(\frac{A_n}{n}\right)^p + \left(\frac{A_{n-1}}{n-1}\right)^p\right]$$

$$= \frac{1}{p-1}\left[\left(\frac{A_n}{n}\right)^p (p-1-np) + (n-1)(p-1) + (n-1)\left(\frac{A_{n-1}}{n-1}\right)^p\right]$$

$$= \frac{1}{p-1}\left[(n-1)\left(\frac{A_{n-1}}{n-1}\right)^p - n\left(\frac{A_n}{n}\right)^p\right].$$

Thus

$$\sum_{n=1}^{m}\left[\left(\frac{A_n}{n}\right)^p - \frac{p}{p-1}\left(\frac{A_n}{n}\right)^{p-1} a_n\right] \leq \frac{-m}{p-1}\left(\frac{A_m}{m}\right)^p \leq 0$$

(the sum telescopes!). Thus, by Hölders Inequality

$$\sum_{n=1}^{m} \left(\frac{A_n}{n}\right)^p \leq \frac{p}{p-1} \sum_{n=1}^{m} \left(\frac{A_n}{n}\right)^{p-1} a_n$$

$$\leq \frac{p}{p-1} \left(\sum_{n=1}^{m} \left(\frac{A_n}{n}\right)^{(p-1)q}\right)^{1/q} \left(\sum_{n=1}^{m} a_n^p\right)^{1/p}$$

$$= \frac{p}{p-1} \left(\sum_{n=1}^{m} \left(\frac{A_n}{n}\right)^p\right)^{1/q} \left(\sum_{n=1}^{m} a_n^p\right)^{1/p}.$$

Thus

$$\left(\sum_{n=1}^{m} \left(\frac{A_n}{n}\right)^p\right)^{1-1/q} \leq \frac{p}{p-1} \left(\sum_{n=1}^{m} a_n^p\right)^{1/p}.$$

But $1 - \frac{1}{q} = \frac{1}{p}$ and Hardy's inequality is proved.

The above is Hardy's original proof somewhat expanded.

XI.4 Theorem. (König) *Let $0 < p < 1$. For $T \in \mathcal{S}_1(H)$ let*

$$s_1^p(T) = \sum_{n=1}^{\infty} \left[\frac{1}{n} \sum_{k=1}^{n} \sigma_k(T)^p\right]^{1/p}.$$

Then

1) $s_1(T) \leq s_1^{(p)}(T) \leq \left[\frac{1}{1-p}\right]^{1/p} s_1(T)$; *and if $S \in \mathcal{S}_1(H)$,*

2) $s_1^p(S + T) \leq \left[\frac{1}{1-p}\right]^{1/p} [s_1^p(T) + s_1^p(S)]$.

Thus, even though s_1^p is not a norm, convergence is the same with respect to s_1 and s_1^p.

Proof: 1) Since $\sigma_1(T) \geq \sigma_2(T) \geq \cdots$ we have

$$\sum_{k=1}^{n} \sigma_k(T)^p \geq n\sigma_n(T)^p.$$

Thus

$$s_1^p(T) = \sum_{n=1}^{\infty} \left[\frac{1}{n} \sum_{k=1}^{n} \sigma_k(T)^p\right]^{1/p} \geq \sum_{n=1}^{\infty} \left(\sigma_n(T)^p\right)^{1/p} = s_1(T).$$

Let $q = \frac{1}{p} > 1$.

$$\sum_{n=1}^{\infty} \left[\frac{1}{n} \sum_{j=1}^{n} \sigma_j(T)^p \right]^{1/p} = \sum_{n=1}^{\infty} \left[\frac{1}{n} \sum_{j=1}^{n} \sigma_j(T)^{1/q} \right]^q$$

$$\leq \left(\frac{q}{q-1} \right)^q \sum_{n=1}^{\infty} \left(\sigma_n(T)^{1/q} \right)^q$$

$$= \left(\frac{q}{q-1} \right)^q s_1(T).$$

This last inequality by XI.3. But $\left[\frac{q}{q-1} \right]^q = \left[\frac{1}{1-p} \right]^{1/p}$ so 1) holds.

2) By part 1)

$$s_1^p(T+S) \leq \left[\frac{1}{1-p} \right]^{1/p} [s_1(S+T)] \leq \left[\frac{1}{1-p} \right]^{1/p} [s_1(S) + s_1(T)]$$

$$\leq \left[\frac{1}{1-p} \right]^{1/p} [s_1^p(S) + s_1^p(T)].$$

We next show how to "localize" an eigenvalue by averaging the singular numbers $\sigma_n(T)$. To simplify notation we write $\beta_{n,p}(T) = \left[\frac{1}{n} \sum_{i=1}^{n} \sigma_i(T)^p \right]^{1/p}$.

Exercise 2.   Let $p > 0$ show that

$$\max \left\{ \sum_{i=1}^{n} x_i : x_i \geq 0 : \sum_{i=1}^{n} x_i^p = 1 \right\} = n^{1-1/p}.$$

XI.5 Lemma.   (Localization of Eigenvalues). Let $T \in K(H)$. Then for any n and $p > 0$, $|\lambda_n(T)| \leq \beta_{n,p}$.

Proof: We are assuming, of course that $|\lambda_1(T)| \geq |\lambda_2(T)| \geq \cdots$. Thus, $n|\lambda_n(T)| \leq \sum_{i=1}^{n} |\lambda_i(T)| \leq \sum_{i=1}^{n} \sigma_i(T)$ by the weak Weyl inequality. But

$$\sum_{i=1}^{n} \sigma_i(T) \leq n^{1-1/p} \left(\sum_{i=1}^{n} \sigma_i(T)^p \right)^{1/p}$$

by exercise 2. Thus

$$|\lambda_n(T)| \leq n^{-1/p} \left(\sum_{i=1}^{n} \sigma_i(T)^p \right)^{1/p} = \beta_{n,p}(T).$$

We need two more properties of $\beta_{n,p}(T)$ before proving the key result of König which leads directly to the trace theorem.

First, the last of the "Lagrange" type inequalities.

Exercise 3. (i) Prove for $0 < p < 1$,

$$\sup\{(x^p + y^p)^{1/p} : x > 0,\ y > 0,\ x + y = 1\} = 2^{1/p-1}.$$

(ii) For $0 < p < 1$ show also that $(a + b)^p \le a^p + b^p$.

We also need the monotonicity of the singular numbers σ_n.

Exercise 4. For $S, T \in K(H)$ and any n, show that

$$\sigma_{2n-1}(T + S) \le \sigma_n(T) + \sigma_n(S).$$

(See Chapter VIII Problem 1 in Notes, Exercises and Hints).

XI.6 Lemma. Let $0 < p < 1$ and $S, T \in K(H)$. Then

(i) $\beta_{2n,p}(S + T) \le 2^{1/p-1}[\beta_{n,p}(S) + \beta_{n,p}(T)]$; and,

(ii) $\beta_{2n-1,p}(T) \le 2^{1/p}\beta_{2n,p}(T)$.

Proof:

$$\beta_{2n,p}(T + S) = \left(\frac{1}{2n} \sum_{i=1}^{2n} \sigma_i(T+S)^p\right)^{1/p}$$

$$= \left(\left(\frac{1}{2n}\right)\left[\sum_{i=1}^{n} \sigma_{2i-1}(T+S)^p + \sum_{i=1}^{n} \sigma_{2i}(T+S)^p\right]\right)^{1/p}$$

$$\le \left(\frac{1}{2n} \cdot 2 \sum_{i=1}^{n} \sigma_{2i-1}(T+S)^p\right)^{1/p} \le \left(\frac{1}{n} \sum_{i=1}^{n} [\sigma_i(S) + \sigma_i(T)]^p\right)^{1/p}$$

$$\le \left(\frac{1}{n} \sum_{i=1}^{n} \sigma_i(T)^p + \frac{1}{n} \sum_{i=1}^{n} \sigma_i(S)^p\right)^{1/p} \le 2^{1/p-1}[\beta_{n,p}(T) + \beta_{n,p}(S)]$$

by exercises 3 and 4.

For (ii) observe that

$$\beta_{2n-1,p}(T) = \left(\frac{1}{2n-1} \sum_{i=1}^{2n-1} \sigma_i(T)^p\right)^{1/p}$$

$$= \left(\left[\frac{1}{2n(2n-1)} + \frac{1}{2n} \right] \sum_{i=1}^{2n-1} \sigma_i(T)^p \right)^{1/p}$$

$$\leq \left[\frac{2}{2n} \sum_{i=1}^{2n} \sigma_i(T)^p \right]^{1/p} = 2^{1/p} \beta_{2n,p}(T).$$

The trouble with $\sigma - tr(T)$ is that, a priori, there appears to be no control on the sums of eigenvalues. König's result below remedies this. His result says that $\mathcal{S}_1$-convergent sequences have eigenvalues sequences with uniformly small "tails" in the ℓ_1-norm. We make this precise.

XI.7 Theorem (König). *Let* $T_0, T_n \in \mathcal{S}_1(H)$, $\lim s_1(T_n - T_0) = 0$ *and* $\varepsilon > 0$. *There is an* N_0 *such that*

$$\sum_{n=N_0}^{\infty} |\lambda_n(T_K)| \leq \varepsilon \quad \text{for all} \quad K \geq 0.$$

Proof: Fix $0 < p < 1$. Choose n_0, m_0 with $s_1^{(p)}(T_0 - T_n) < \varepsilon_p$ for $n \geq n_0$ and $\sum_{m=m_0}^{\infty} \beta_{m,p}(T_0) < \varepsilon_p$ (possible by XI.4.) By the localization of eigenvalues lemma (XI.5) $|\lambda_K(T_n)| \leq \beta_{K,p}(T_n)$ for any K. Thus by XI.6 we have

$$\sum_{m=2m_0}^{\infty} |\lambda_m(T_n)| \leq \sum_{m=2m_0}^{\infty} \beta_{m,p}(T_n) = \sum_{m=m_0}^{\infty} \beta_{2m-1,p}(T_n) + \sum_{m=m_0}^{\infty} \beta_{2m,p}(T_n)$$

$$\leq \left(2^{1/p} + 1 \right) \sum_{m=m_0}^{\infty} \beta_{2m,p}(T_n) = \left[2^{1/p} + 1 \right] \sum_{m=m_0}^{\infty} \beta_{2m,p}(T_0 - [T_0 - T_n])$$

$$\leq \left[2^{1/p} + 1 \right] 2^{1/p-1} \left[\sum_{m=m_0}^{\infty} \beta_{m,p}(T_0) + \sum_{m=m_0}^{\infty} \beta_{m,p}(T_0 - T_n) \right]$$

$$\leq \left[2^{1/p} + 1 \right] 2^{1/p-1} (\varepsilon_p + \varepsilon_p) < \varepsilon \quad \text{for suitable} \quad \varepsilon_p,$$

and this holds for all $n \geq n_0$. By making m_0 larger if necessary we can handle the $T_1, \ldots, T_{n_0}$ in the same manner and so the theorem is proved.

Our final goal is to show that $\sigma - tr(T)$ is continuous on $\mathcal{S}_1(H)$. Perhaps a brief glance at chapter V to refresh your memory is in order here.

Exercise 5. If $T \in \mathfrak{L}(H)$ and C is a circle contained in $\rho(T)$, $P = \frac{1}{2\pi i} \int_C R_\zeta d\zeta$ is a projection ($P^2 = P$), and also $TP = PT$.

[HINT] Let D be a circle completely inside C. Then
$\int_C R_\zeta d\zeta = \int_D R_\zeta d\zeta$ (why?). Then $(2\pi i)^2 P^2 = \int_C R_\zeta d\zeta \cdot \int_D R\eta d\eta = \int_C \int_D (R\eta - R_\zeta)(\zeta - \eta)^{-1} d\zeta d\eta = (2\pi i)^2 P$ (why?).

Since P is a projection, P is the identity on H_P, the range of P. Since $TP = PT$, the restriction T_1 of T to H_P is in $\mathcal{L}(H_P)$. If $Q = I - P$ and H_Q denotes the range of Q, T_2, the restriction of T to H_Q, is in $\mathcal{L}(H_Q)$. Moreover $H = H_P \oplus H_Q$. Finally, since each R_ζ, $\zeta \in \rho(T)$, commutes with T, each such R_ζ commutes with P and Q. We let R_ζ^1 and R_ζ^2 denote the restrictions of R_ζ to H_P and H_Q respectively.

Exercise 6. Show that R_ζ^1 and R_ζ^2 are the resolvent operators of T_1 and T_2 respectively.

XI.8 Theorem. *For $T \in \mathcal{L}(H)$*

$$\rho(T) = \rho(T_1) \cap \rho(T_2).$$

Proof: Exercise 6 proves part of the assertion. Suppose $\zeta \in \rho(T_1) \cap \rho(T_2)$. Then there exists operators S_1 and S_2 such that

$$S_1(\zeta I_P - T_1) = I_P \quad \text{and} \quad S_2(\zeta I_Q - T_2) = I_Q.$$

Here I_P and I_Q denote the identity operator on H_P and H_Q respectively. Let $S = S_1 \oplus S_2$, i.e.,

$$S(x) = S_1(Px) + S_2(Qx).$$

Then $S(\zeta I - T)x = x$ for each x in H_P and each $x \in H_Q$. Since $H = H_P \oplus H_Q$, $\zeta \in \rho(T)$ and $S = R_\zeta$ so $S_1 = R_\zeta^1$ and $S_2 = R_\zeta^2$.

Stated in a complementary fashion, $\zeta \in \sigma(T)$ if and only if

$$\zeta \in \sigma(T_1) \cup \sigma(T_2).$$

That is, the projection P splits the spectrum of T into two parts. What are those two sets?

XI.9 Theorem. *With T_1 and T_2 having the meanings above and int C, ext C denoting the interior and exterior of C, $\sigma(T_1) = \sigma(T) \cap \text{int } C$ and $\sigma(T_2) = \sigma(T) \cap \text{ext } C$.*

Proof: First we show that if $\zeta \in \text{ext } C$ then $\zeta \in \rho(T_1)$. Suppose $\eta \in \rho(T)$. Then $(\zeta I - T)R_\eta = (\eta I - T)R_\eta + (\zeta - \eta)R_\eta$. Thus

$$(\zeta I - T)\circ \frac{1}{2\pi i} \int_C R_\eta(\zeta - \eta)^{-1} d\eta$$

$$= \left(\frac{1}{2\pi i} \int_C (\zeta - \eta)^{-1} d\eta \right) I + \frac{1}{2\pi i} \int_C R_\eta d\eta = P.$$

(Why?) But P is the identity on H_P so $(\zeta I_P - T_1)^{-1}$ exists. That is $\zeta \in \rho(T_1)$ and so $\sigma(T_1) \subset \text{int } C$. However we have just observed that $\sigma(T_1) \subset \sigma(T)$ and so the first assertion is proved.

If $\zeta \in \text{int } C$, by the same trick just used,

$$(\zeta I - T_2)\circ \frac{1}{2\pi i} \int_C R_\eta(\zeta - \eta)^{-1} d\eta = -I + P = -Q.$$

Thus $(\zeta I_Q - T_2)^{-1}$ exists.

For $T \in K(H)$ and $r > 0$ we will say that r is *T-admissible* provided the circle $C_r = \{\lambda \in \mathbb{C} : |\lambda| = r\}$ is in the resolvent, $\rho(T)$, of T. We will write $P_r = \frac{1}{2\pi i} \int_{C_r} R_\lambda d\lambda$. Recall that P_r is a projection ($P_r^2 = P_r$). If I denotes the identity operator on H we write $T^i = TP_r$ and $T^e = T(I - P_r)$.

Exercise 7. a) Show that $\sigma(T^i) = \{\lambda \in \sigma(T) : |\lambda| < r\}$ and $\sigma(T^e) = \{\lambda \in \sigma(T) : |\lambda| > r\}$. Thus the i, interior and e, exterior.

b) If $T \in K(H)$ and $r > 0$ is T-admissible show that rank $T^e < +\infty$. [HINT] Recall that $\sigma(T) = \{\lambda_n\} \cup \{0\}$ and $\lim_{n \to \infty} \lambda_n = 0$ in this case. Use (a).

c) If $T \in \mathcal{S}_1(H)$ (so, in particular, $\sum \lambda_n(T)$ exists by the weak Weyl inequality) then

$$\sigma - \text{tr}(T) = \sigma - \text{tr}(T^i) + \sigma - \text{tr}(T^e).$$

XI.10 Main Theorem. *The function $\sigma - \text{tr}(T)$ is continuous on $\mathcal{S}_1(H)$.*

Proof: (Leiterer and Pietsch): Let $T_n, T_0 \in \mathcal{S}_1(H)$, $n \geq 1$ with $\lim_{n \to \infty} s_1(T_n - T_0) = 0$. By XI.7 for $\varepsilon > 0$ there is an N such that

$$\sum_{n=N+1}^{\infty} |\lambda_n(T_j)| \leq \varepsilon \quad \text{for} \quad j = 0, 1, 2, \dots.$$

Choose a T_0-admissible $r > 0$ with $Nr < \varepsilon$ (possible since $\sigma(T_0) = \{\lambda_n(T_0)\} \cup \{0\}$ where $\sum |\lambda_n(T_0)| < +\infty$ — use exercise 1(a)). Also by exercise 1(a), $(\lambda_n(T_0^i))$ consists of those $\lambda_n(T_0)$ satisfying $|\lambda_n(T_0)| < r$. Thus

$$\sigma - \text{tr}(T_0^i) \leq \sum_{n=1}^{N} |\lambda_n(T_0^i)| + \sum_{n=N+1}^{\infty} |\lambda_n(T_0^i)| < 2\varepsilon.$$

Let $c = \max\{\|R_\lambda\| : |\lambda| = r\}$ and let q, $0 < q < 1$ satisfy $\frac{q}{c} + \frac{cqr^2}{1-q} < \varepsilon$. (Recall that $R_\lambda = (\lambda I - T_0)^{-1}$ and $\|R_\lambda\|$ is continuous on C_r. Thus c exists and is clearly greater than 0. Moreover c depends only on r. Thus q depends only on r and ε.)

We break the remainder of the proof into a number of claims. Choose j_0 such that $j \geq j_0$ implies $s_1(T_j - T_0) < q/c$.

Claim 1. r is T_j-admissible for all $j \geq j_0$. Indeed if $|\lambda| = r$ let

$$S = R_\lambda \left[I + \sum_{n=1}^{\infty} [T_j - T_0]^n R_\lambda^n \right].$$

Then $\|S\| \leq \|R_\lambda\| + \|R_\lambda\| \sum_{n=1}^{\infty} \left(\frac{q}{c}\right)^n c^n < +\infty$ since $0 < q < 1$.

Moreover, by the Neumann expansion

$$I + \sum_{n=1}^{\infty} ([T_j - T_0] R_\lambda)^n = [I(T_j - T_0)R_\lambda]^{-1}.$$

Thus

$$\begin{aligned}
(\lambda I - T_j)S &= ([\lambda I - T_0] - (T_j - T_0))S \\
&= \left[R_\lambda^{-1} - (T_j - T_0) \right] \left[R_\lambda \left[I - (T_j - T_0)R_\lambda \right]^{-1} \right] \\
&= \left[I - (T_j - T_0)R_\lambda \right]^{-1} \left[I - (T_j - T_0)R_\lambda \right]^{-1} = I
\end{aligned}$$

so $S = R_\lambda(T_j)$ which we denote by $R_\lambda(j)$.

Claim 2. $s_1(R_\lambda(j) - R_\lambda) \leq \frac{cq}{1-q}$ for any $j \geq j_0$. To see this, by what was just shown

$$\begin{aligned}
s_1(R_\lambda(j) - R_\lambda) &= s_1 \left(R_\lambda \sum_{n=1}^{\infty} [T_j - T_0]^n R_\lambda^n \right) \leq \|R_\lambda\| s_1 \left(\sum_{n=1}^{\infty} [T_j - T_0]^n R_\lambda^n \right) \\
&\leq \|R_\lambda\| \sum_{n=1}^{\infty} s_1(T_j - T_0)^n \|R_\lambda\|^n \leq c \sum_{n=1}^{\infty} \left(\frac{q}{c}\right)^n c^n = c \sum_{n=1}^{\infty} q^n = \frac{cq}{1-q}
\end{aligned}$$

since $q < 1$.

Claim 3. $T^i = TP_r = \frac{1}{2\pi i}\int_{C_r}\lambda R_\lambda d\lambda$. Indeed,

$$T^i = TP_r = T \circ \frac{1}{2\pi i}\int_{C_r}R_\lambda d\lambda.$$

But

$$\frac{1}{2\pi i}\int_{C_r}(T - \lambda I)R_\lambda d\lambda + \frac{1}{2\pi i}\int_{C_r}\lambda R_\lambda d\lambda = 0 + \frac{1}{2\pi i}\int_{C_r}\lambda R_\lambda d\lambda,$$

since

$$I\int_{C_r}d\lambda = I\int_0^{2\pi}rie^{i\theta}d\theta = 0.$$

Likewise, $T_j^i = \frac{1}{2\pi i}\int_{C_r}\lambda R_\lambda(j)d\lambda$.

Claim 4. $s_1(T_0^i - T_j^i) \le \frac{cr^2 q}{1-q}$. Indeed,

$$s_1(T_0^i - T_j^i) = \frac{1}{2\pi}s_1\left[\int_{C_r}\lambda[R_\lambda - R_\lambda(j)]d\lambda\right] \le \frac{r}{2\pi}\int_{C_r}s_1\left(R_\lambda - R_\lambda(j)\right)|d\lambda|$$

$$\le \frac{r}{2\pi}\int_{C_r}\frac{cq}{1-q}|d\lambda|$$

by claim 2. Thus $s_1(T_0^i - T_j^i) \le \frac{cr^2 q}{1-q}$ since $\int_{C_r}|d\lambda| = \int_0^{2\pi}rd\theta = 2\pi r$.

Claim 5. $s_1(T_j^e - T^e) \le \frac{q}{c} + \frac{cr^2 q}{1-q}$. This is because

$$s_1(T_j^e - T_0^e) = s_1\left([T_j - T_j^i] - [T_0 - T_0^i]\right) \le s_1(T_j - T_0) + s_1(T_j^i - T_0^i)$$

$$\le \frac{q}{c} + \frac{cr^2 q}{1-q}.$$

Now we are ready to roll: T_j^e and T_0^e are finite rank so

$$|\sigma - \text{tr}(T_j^e) - \sigma - \text{tr}(T_0^e)| = |\sigma - \text{tr}(T_j^e - T_0^e)| \le s_1(T_j^e - T^e)$$

$$\le \frac{q}{c} + \frac{cr^2 q}{1-q} < \varepsilon.$$

The next to last inequality coming from the weak Weyl inequality. Thus

$$|\sigma - \text{tr } T_j - \sigma - \text{tr } T_0| = |\sigma - \text{tr}(T_0^i - T_j^i) - \sigma - \text{tr}(T_0^e - T_0^e)|$$

$$\le |\sigma - \text{tr}(T_j^e)| + |\sigma - \text{tr}(T_0^e)| + |\sigma - \text{tr}(T_j^i - T_0^i)|$$

$$\le 2\varepsilon + 2\varepsilon + \varepsilon = 5\varepsilon.$$

This is valid since

$$\sum_{j=1}^{N} |\lambda_j(T_j^i)| \leq N\lambda_1(T_j^i) \leq Nr < \varepsilon$$

since r is T_j-admissible. The estimates and the claims yield the other inequalities.

XI.11 (Lidskij Trace Theorem). *For $T \in \mathcal{S}_1(H)$, $\sigma - \text{tr}(T) = \phi - \text{tr}(T)$.*

Proof: We need only to put the pieces together. We have just shown that $\sigma - \text{tr}(T)$ is *continuous* on $\mathcal{S}_1(H)$ (but not necessarily linear!). Thus $\sigma - \text{tr}(T)$ is a continuous extension of $\sigma - \text{tr}$ from the finite rank operators. However that extension is unique and we denoted it by $\widetilde{\text{tr}}$. In XI.2 we showed that $\widetilde{\text{tr}} = \phi - \text{tr}$ on $\mathcal{S}_1(H)$. Moreover $\phi - \text{tr}$ is linear. By uniqueness $\phi - \text{tr}(T) = \sigma - \text{tr}(T)$ on $\mathcal{S}_1(H)$.

We belabor the point but again wish to emphasize the difficulties: an operator $T \in K(H)$ is said to be *quasi-nilpotent* if $\sigma(T) = \{0\}$. Clearly if $T \in \mathcal{S}_1(H)$ is quasi-nilpotent $\sigma - \text{tr}(T) = 0$. If such a T has an *arbitrary* representation

$$Tx = \sum_{n=1}^{\infty}(x, f_n)x_n, \ \sum_{n=1}^{\infty} \|f_n\| \, \|x_n\| < +\infty$$

it is incredible that $\sum_{n=1}^{\infty} x_n(f_n) = 0$. None-the-less it is true.

Also if $T, S \in \mathcal{S}_1(H)$, it is trivial that $\phi - \text{tr}(S+T) = \phi - \text{tr}(S) + \phi - \text{tr}(T)$. If such T, S have eigenvalues $(\lambda_n(T))$, $(\lambda_n(S))$ it is not at all obvious that

$$\sum_{n=1}^{\infty} \lambda_n(S+T) = \sum_{n=1}^{\infty} \lambda_n(S) + \sum_{n=1}^{\infty} \lambda_n(T).$$

None-the-less it is true.

What good is the trace formula? How can it be put to use? This is somewhat difficult to answer at this hopefully elementary level, but one application is fairly easy to discuss: Since the beginnings of Banach space theory one of the major problems has been to describe the conjugate space X^* of a given Banach space X. We have seen (Riesz-Representation theorem) that in some reasonable fashion H^* can be identified with H. In these notes we have introduced the spaces of operators $S_p(H)$. What is the conjugate space $\left(S_p(H)\right)^*$? Writing for now $S_\infty(H) = K(H)$ the answer is $\left(S_p(H)\right)^* = S_q(H)$, $\frac{1}{p} + \frac{1}{q} = 1$, $1 < p < +\infty$, $S_1(H)^* = \mathcal{L}(H)$ and $S_\infty(H)^* = S_1(H)$. Here "=" means "is isometrically isomorphic to". How does $T \in S_q(H)$ act as a linear functional on $S_p(H)$? Well, since $\frac{1}{p} + \frac{1}{q} = 1$ and $T \in S_q(H)$, $S \in S_p(H)$, $ST \in S_1(H)$. Define $T(S) = \text{tr}(ST)$ (either $\sigma - \text{tr}$ or $\phi - \text{tr}$!).

We have just hinted at "trace duality" a still viable area of research in Banach space theory.

One should examine some of the proofs of the Lidskij trace formula in the literature, e.g., in vol II of Dunford and Schwartz, Simon, and, perhaps the translation of Lidskij's original paper - see the given references.

Final Remarks, Exercises and Hints

We mention briefly two results about eigenvalues and singular numbers of operators on Hilbert space.

1. (**Horn**) Prove that if $S, T \in K(H)$ $\prod_{i=1}^{n} \sigma_i(ST) \leq \prod_{i=1}^{n} \sigma_i(S)\sigma_i(T)$.

[HINT] There is nothing to prove if $\sigma_n(ST) = 0$. So suppose $\sigma_n(ST) \neq 0$. Then $(ST)^{\frac{1}{2}} = UST$ where U is a partial isometry. Let H_n be a n-dimensional eigenspace for $\left(\lambda_i(ST)^{\frac{1}{2}}\right)_{i=1}^{n}$. Let P, Q be orthogonal projections onto H_n and $T(H_n)$.

Then

$$\prod_{j=1}^{n} \sigma_j(ST) = \prod_{j=1}^{n} \lambda_j([U\,ST]^{\frac{1}{2}}) = \det(PUSQ)\det(QTP)$$

$$= \prod_{i=1}^{n} \sigma_j(PUSQ)\sigma_j(QTP).$$

See the ideas in Appendix B.

2. [**Ky Fan**] Let $S, T \in K(H)$. Show that

$$\sum_{j=1}^{n} \sigma_j(S+T) \leq \sum_{j=1}^{n} \sigma_j(S) + \sigma_j(T).$$

[HINT] Write the Schmidt decomposition

$(S+T)x = \sum_{j=1}^{\infty} \sigma_j(S+T)(x, x_j)Ux_j$, U a partial isometry, and let P be the orthogonal projection onto $sp\{x_i \cdots x_n\}$. Then

$$\sum_{j=1}^{n} \sigma_j(S+T) = \operatorname{tr}(PU^*(S+T)P)$$

$$= \operatorname{tr}(PU^*SP) + \operatorname{tr}(PU^*TP).$$

Now use the results in this chapter.

This result of Ky Fan shows that the Schatten 1 -class is normed by $s_1(T)$ without resorting to the notion of the nuclear operator.

It is beyond the scope of this work to hint at the proof, but we end with a beautiful result of König which generalizes the spectral radius formula:

Let $T \in K(X)$ where X is an arbitrary Banach space. Then

$$|\lambda_n(T)| = \lim_{r \to \infty} \alpha_n \left(T^r\right)^{\frac{1}{r}}$$

where $(\lambda_n(T))$ is the eigenvalue sequence of T and $\alpha_n(\bullet)$ denotes the approximation numbers.

APPENDIX C: LOCALIZATION OF EIGENVALUES

If one uses the multiplicative Weyl inequality (Appendix B), the localization of eigenvalues lemma (XI.7) is an easy consequence of the inequality (yes, another inequality) comparing the arithmetic and geometric means: For $a_i \geq 0$

$$\left(\prod_{i=1}^{n} a_i \right)^{1/n} \leq \frac{1}{n}(a_1 + a_2 + \cdots + a_n).$$

Indeed,

$$a_1 a_2 = \left(\frac{a_1 + a_2}{2} \right)^2 - \left(\frac{a_1 - a_2}{2} \right)^2 \leq \left(\frac{a_1 + a_2}{2} \right)^2.$$

Repeating (by pairing)

$$a_1 \ldots a_2 m \leq \left(\frac{a_1 + a_2 + \cdots + a_2 m}{2^m} \right)^{2^m}.$$

If $n < 2^m$ let $b_1 = a_1, b_2 = a_2, \ldots, b_n = a_n,$

$$b_{n+1} = b_{n+2} = \cdots = b_{2m} = \frac{a_1 + a_2 + \cdots + a_n}{n}.$$

Then,

$$a_1 a_2 \ldots a_n \left(\frac{a_1 + a_2 + \cdots + a_n}{n} \right)^{2^m - n}$$

$$\leq \left[\left(\frac{n \left(\frac{\sum\limits_{i=1}^{n} a_i}{n} \right) + (2^m - n) \left(\frac{\sum\limits_{i=1}^{n} a_i}{n} \right)}{2^m} \right) \right]^{2^m}$$

$$\leq \left(\frac{\sum\limits_{i=1}^{n} a_i}{n} \right)^{2^m}$$

so

$$a_1 a_2 \ldots a_n \leq \left(\frac{a_1 + a_2 + \cdots + a_n}{n} \right)^{n}.$$

(This is the original argument (and still the best) of Hardy, Littlewood and Polya).

C.1 Localization of Eigenvalues. (See XI.7). If $0 < p < 1$ and $T \in K(H)$

$$|\lambda_n(T)| \leq \left[\frac{1}{n} \sum_{i=1}^{n} \sigma_i(T)^p \right]^{1/p}.$$

Proof: By Weyl's multiplicative inequality

$$\prod_{k=1}^{n} |\lambda_k(T)| \leq \prod_{k=1}^{n} \sigma_k(T).$$

Since $|\lambda_1(T)| \geq |\lambda_2(T)| \geq \cdots$ we obtain

$$|\lambda_n(T)| \leq \left(\prod_{k=1}^{n} |\lambda_k(T)| \right)^{1/n} \leq \frac{1}{n} \left(\sum_{k=1}^{n} \sigma_k(T) \right)$$

by the arithmetic-geometric means result. Since $0 < p < 1$ the result follows.

BIBLIOGRAPHY

Books

1. Ruel Churchill, *Introduction to Complex Variables and Applications*, McGraw-Hill, New York 1948 (or any subsequent edition).

2. Nelson Dunford, J. T. Schwartz, *Linear Operators I and II*, Interscience, New York, 1958 and 1973.

3. Paul Halmos, *Introduction to Hilbert Space and the Theory of Spectral Multiplicity*, Chelsea, New York, 1951.

4. G. H. Hardy, J. E. Littlewood, G. Polya, *Inequalities*, Cambridge Univ. Press, 1934.

5. Edgar Lorch, *Spectral Theory*, Oxford Univ. Press, 1962.

6. Albrecht Pietsch, *Operator Ideals*, North Holland, 1980.

7. F. Riesz, B. Sz. Nagy, *Functional Analysis*, Ungar 1955.

8. Barry Simon, *Trace Ideals and their applications*, Cambridge Univ. Press, 1979.

9. A. E. Taylor, *Introduction to Functional Analysis*, John Wiley and Sons, New York, 1958.

Research Papers

1. Hermann König, s-numbers, eigenvalues and the trace theorem in Banach spaces, *Studia Math.* **67** (1980), pp. 157-1872.

2. H. Lieterer, A. Pietsch, An elementary proof of Lidskij's trace formula, *Wiss. Zeitschr. Univ. Jena* (1982), pp. 587–594.

3. V. B. Lidskij, Non-self adjoint operators with a trace, *Dokl. Akad. Nauk* (1959), pp. 485–487 (Russian).

Future Reading

1. Hermann König, *Eigenvalue Distribution of Compact Operators*, Birkhauser, Stuttgart, 1986.

2. Albretsch Pietsch, Eigenvalues and s-numbers, *Geest and Partig K-G*, Leipzig, 1987 (also published by Cambridge Univ. Press).

INDEX OF NOTATION

INDEX OF TERMS